WINGS OF MADNESS

PAUL HOFFMAN

WINGS OF MADNESS

ALBERTO SANTOS-DUMONT AND THE INVENTION OF FLIGHT

FOURTH ESTATE • *London* and *New York*

Photo credits:
1. Courtesy Carmen Rossignoli. Scanning courtesy of Stella Villares Guimarães.
2–4, 6–13, 15, 17, 18, 21, 22, 24–36. From the collection of General Nelson Wanderley. Courtesy Sophia Helena Dodsworth Wanderley. Scanning courtesy of Stella Villares Guimarães.
5. Courtesy Cartier.
14, 16, 19, 20. Reproduced from *My Air-Ships,* 1904.
23. Courtesy Bibliothèque Forney.

First published in Great Britain in 2003 by
Fourth Estate
A Division of HarperCollins*Publishers*
77–85 Fulham Palace Road,
London w6 8jb
www.4thestate.com

1 3 5 7 9 10 8 6 4 2

A catalogue record for this book is available from the
British Library

isbn 1–84115–368–0

Printed in Great Britain by
Clays Ltd, St Ives plc

For Ann and Alexander and Matt

Contents

Contents] ix

WINGS OF MADNESS

●

IN DECEMBER 1903, an eleven-year resident of Paris, the Brazilian aeronautical pioneer Alberto Santos-Dumont, held a small holiday party in his high-ceilinged apartment on the Champs-Elysées. Louis Cartier, the jeweler, was there, as was Princess Isabel, the daughter of the last emperor of Brazil. The other attendees can only be surmised because there was no printed guest list, but his regular dining partners and confidants included George Goursat, the flamboyant writer and cartoonist who drew caricatures of the rich and famous on the walls of the city's fanciest restaurants; Gustave Eiffel, the architect of the eponymous tower; Antônio Prado Jr., the son of the Brazilian ambassador; two or three Rothschilds, who first met their thirty-year-old host when his experimental airship crashed in their gardens; Empress Eugénie, Napoleon III's reclusive widow; and assorted kings, queens, dukes, and duchesses too numerous to name.

When Santos-Dumont's butler ushered the guests into the dining room, they were amused to find that they had to climb a stepladder so that they could sit on tall chairs positioned around a table higher than they were. But they were not surprised. Since the late 1890s Santos-Dumont had been giving "aerial dinner parties." The first ones were held at an ordinary table and chairs suspended by wire from the ceiling. This worked when the hundred-pound Santos-Dumont dined alone, but when a group assembled, the ceiling eventually gave way under the collective weight. Santos-Dumont was a skilled craftsman, who had learned woodworking from the men on his father's coffee plantation, so he built the long-legged tables and chairs that had become a fixture of his apartment ever since. At the first elevated soirees, his guests, between sips of milky green absinthe, invariably asked what the point of the high table was. And their shy host, who preferred to let others do the talking, would run his bejeweled fingers through his jet-black hair, which was parted in the middle, in a style seen almost exclusively on women, and impishly explain that they were dining aloft so that they could imagine what life was like in a flying machine. The guests laughed. Flying machines did not exist in the 1890s, and received scientific wisdom said that they never would. Santos-Dumont ignored the snickering and insisted that they would soon be commonplace.

Hot-air balloons, to be sure, were a familiar sight in the skies of fin de siècle Paris, but they were not flying machines. With no source of power, these large floating orbs—they were described as spherical although they actually had the shape of an inverted pear—were entirely at the mercy of the wind. By the turn of the century, Santos-Dumont changed that. He strapped an automobile engine and propeller to the balloon and, to make it aerodynamically efficient, switched its shape to that of a sleek cigar. On October 19, 1901, thousands of people turned out to watch him circle the Eiffel Tower in his innovative airship. The crowds on the bridges over the Seine were so thick that people were shoved into the river when they scaled the parapets to get a better view. The scientists who observed the flight from Gustave Eiffel's apartment at the top of the tower were sure he would not make it. They feared that an unpredictable wind would impale him on the spire. Others were convinced that the balloon would explode. When Santos-Dumont proved them wrong, Jules Verne and H. G. Wells sent congratulatory telegrams.

By the end of 1903, at the time of Santos-Dumont's dinner with Cartier and Princess Isabel, he was a fixture in the Paris skies. He had designed a small airship, which his fans called *Baladeuse* ("Wanderer"), his personal runabout in which he went barhopping, tying the balloon to the gas-lamp posts in front of the city's glamorous night spots. *Baladeuse* was as easy to operate as that new invention, the automobile, that sputtered down Paris boulevards but it had the advantage of not startling horses or pedestrians when it was in midflight. Santos-Dumont's larger racing airships demanded more attention than *Baladeuse*, and he complained to Cartier that he could not time his own flights because it was dangerous for him to take his hands off the controls and fish out his pocket watch. Cartier promised he would come up with a solution, and he soon invented one of the first wristwatches for Santos-Dumont—a commercial version of which became a must-have accessory for status-conscious Parisians.

Santos-Dumont had a romantic vision of every person on earth possessing their own *Baladeuse,* so that they would literally be free as birds to travel anywhere they wanted anytime they pleased. The future of flying machines, he thought, lay in the lighter-than-air balloon not in the heavier-than-air plane, which as far as he knew had not progressed beyond the unpowered glider. He envisaged gigantic airships—not rigid zeppelins but big, soft balloons with their payloads slung below—

whisking travelers between Paris and New York, Berlin and Calcutta, Moscow and Rio de Janeiro.

Santos-Dumont did not believe in patents. He made the blueprints of his airships freely available to anyone who wanted them. He saw the flying machine as a chariot of peace, bringing estranged cultures in contact with one another so that they could get to know one another as people, thereby reducing the potential for hostilities. In retrospect it seems a naive vision, with the Great War only a decade away, but his optimism was not uncommon among men of science at the turn of the century, when novelties such as the electric light, the automobile, and the telephone were transforming society in fundamental ways.

That December night in 1903, Santos-Dumont and his elevated companions reflected on what a great year it had been for him. He had had none of the usual accidents, which had made him famous as the man who defied death time and again. None of his customary crashes on the jagged rooftops of Parisian hotels, no unexpected nosedives into the Mediterranean, no sudden descents onto a stranger's land. It had been a tranquil year. In *Baladeuse* he owned the skies of France. He was the only one who was consistently puttering around in a flying machine. As Santos-Dumont's butler decanted wine for the guests, Cartier and Princess Isabel offered a toast to their host's ingenuity. No one else was close to mastering the air—or so it seemed.

Eager for a new challenge, Santos-Dumont joined the competition to build and fly the world's first airplane. For a few months he appeared to have succeeded, but after an acrimonious priority fight, Wilbur and Orville Wright, who had initially flown in near secrecy, garnered that glory. Santos-Dumont retained the distinction of flying the first airplane in

Europe, and his élan and perseverance were credited with inspiring aeronauts across the continent.

Early aeronautics in Europe had the quality of a gentlemen's club. Balloon meets on Sunday mornings replaced polo and fox hunts. Flying machines were a diversion for the rich men who owned the first automobiles—oil barons, well-heeled lawyers, and newspaper tycoons. They accepted Santos-Dumont as one of their own because he was the fine-mannered son of a coffee magnate. They supported the inventors of dirigibles and airplanes both by funding them directly and by offering lucrative prizes for aeronautical "firsts": the first to circumvent the Eiffel Tower in a powered balloon, the first to fly an airplane fifty yards, the first to cross the English Channel.

The recreational aspect of these aeronautical contests tended to belie how dangerous they were. More than two hundred men, many of them with wives and children, some of them the top engineers and inventors of their day, died in accidents before Santos-Dumont succeeded. The aeronautical pioneers had none of the modern techniques for assessing the airworthiness of a flying machine. The only way to demonstrate that it could fly was to go up in it, and, as it turned out, most of these fanciful machines either could not get off the ground, stay upright in the air, or descend safely. Santos-Dumont clearly knew the risks involved. And although he told friends that flying gave him the greatest pleasure in life, he would not have courted danger if it were not for a higher purpose—the invention of a technology that would revolutionize transportation and advance world peace.

The first half of his goal was realized in his lifetime. The flying machine of course is now the principal means of conveying people long distances. In the United States alone, there are 90,700 plane flights a day. And in Brazil 157 planes depart for Europe every week. The flight from São Paulo to Paris is eleven hours, a journey that took Santos-Dumont more than a week by steamship and train. Progress toward the second half of his goal has been decidedly mixed. On the one hand, passenger planes, along with the telephone, radio, television, and now the Internet, have turned the world into a global village. When an earthquake strikes El Salvador, food from London can be airlifted there within hours. When an Ebola outbreak is detected in the Congo, doctors from the Centers for Disease Control can be there in a day. On the other hand, military aircraft have caused millions of casualties not just at Hiroshima and Nagasaki but in the ordinary course of war. And then on the morning of September 11, 2001, the unthinkable happened: Two passenger planes were diabolically converted into skyscraper-obliterating missiles. The first great invention of the twentieth century had become the nightmare of the twenty-first.

The Wright brothers had a different motivation from Santos-Dumont in developing the plane. They were not idealists. They did not dream about bringing distant peoples together. They were not thrill-seekers. They did not rhapsodize about the joys of flying or preach a kind of aerial spirituality. They were not playful men and certainly did not host dinners at high tables. They were intent on building flying machines for financial gain, and when the U.S. government initially refused to fund them, they had no moral compunctions about approaching foreign militaries.

In the aftermath of the Great War, when it was evident that the plane could be used as a weapon of mass destruction, Santos-Dumont was the first aeronaut to press for the demilitarization of aircraft. His was a lonely voice, calling on heads of state to decommission their bombers. Orville Wright did not join his call (and Wilbur by then was dead).

Santos-Dumont was perhaps the most revered man in Paris in the first years of the twentieth century. His dapper countenance stared out from cigar boxes, matchbooks, and dinner plates. Fashion designers did a brisk business with replicas of his trademark panama hat and the stiff, high shirt collars that he favored. Toy makers could not turn out enough models of his balloons. Even French bakers honored him, offering cigar-shaped pastries decorated in the colors of the Brazilian flag.

He was famous on both sides of the English Channel—indeed, on both sides of the Atlantic. "When the names of those who have occupied outstanding positions in the world have been forgotten," the London *Times* declared in 1901, "there will be a name which will remain in our memory, that of Santos-Dumont."

The irony in the *Times*'s statement of course is that today he is barely remembered outside Brazil, where he is still a hero of mythic proportion. A town, a major airport, and dozens of streets are named for him. The mere mention of his name brings a smile to most Brazilians, as they picture the bygone era when their daring countryman proudly ruled the skies in a tiny balloon. As the rest of the world has largely forgotten Santos-Dumont, Brazilians themselves, in their romanticizing of the man in poems, songs, statues, busts, paintings, biographies, and memorial celebrations, have neglected his darker side. He was a tortured genius, a free spirit who strove to escape the confines of gravity, the peer pressure of his aeronautical confreres, the isolation of his rural upbringing, the small-mindedness of science's ruling elders, the conformity of married life, the stereotypes of gender, and even the fate of his own cherished invention.

Many boys have dreamed of owning a personal flying machine, a kind of winged car that could take off and land anywhere, without the need of a runway. No one in the twenty-first century has realized that dream. A few elite corporate moguls have come close: They commute to work by helicopter, flying between backyard landing pads and office rooftops. But even a globe-trotting captain of industry cannot fly to his favorite restaurant, the theater, or the store. Only one man in history has enjoyed that freedom. His name was Alberto Santos-Dumont, and his aerial steed was an engine-driven balloon.

[CHAPTER 1]

A R R I V A L

M I N A S   G E R A I S ,   1 8 7 3

•

IN THE LATE eighteenth century, the professional class in Brazil was chafing after three hundred years of Portuguese rule. Deprived of books and newspapers, because the royal family back in Lisbon did not want them to acquire rebellious ideas, the Brazilian colonists nonetheless learned about the American Revolution and the egalitarian philosophy of the French Enlightenment. In 1789, a dentist and second lieutenant in the army named Joaquim José da Silva Xavier, better known by his nickname Tiradentes ("tooth-puller"), helped organize Brazil's first independence movement, *Inconfidência Mineira*, and conspired with other army officers, gold-mine owners, priests, and lawyers to oust the Visconde of Barbacena, the Portuguese representative who governed the state of Minas Gerais. An informer within the movement tipped off the authorities, and Tiradentes was arrested before he could unseat the governor. To discourage other potential fomenters, the authorities hung the dentist in public, hacked his body to pieces, and propped his head and other organs on prominent signposts, all the while proclaiming their loyalty to the queen of Portugal, Dona

Maria, also known as the Crazy One because of her incapacitating melancholy.

After making an example of Tiradentes, the Portuguese crown was lulled into a false sense of security that insurrection was no longer possible in Brazil. The royal family had more pressing things on their minds. Napoleon Bonaparte was on the warpath in Western Europe. In 1807, his forces moved into Portugal, the one country that was still a leak in his European-wide trade blockade against his enemy Great Britain. Prince Dom João VI, who had ruled Portugal since 1792, when his mother was officially declared insane, decided to get out of harm's way by shifting the entire royal court from Lisbon to Rio de Janeiro. It was the first time a European monarchy set up court in the New World. A convoy of ships escorted by the British navy transported the royal family and some ten thousand top Portuguese minds—Supreme Court justices, bankers, clergymen, doctors, and a surgeon named Joaquim José dos Santos. The man was Santos-Dumont's maternal grandfather.

When the Crazy One died in 1816, Dom João assumed the throne. He encouraged immigration to Brazil not just from Portugal but from Spain, France, and Britain. He made the country attractive to professionals by lifting the ban on reading material. He opened theaters and libraries and established scientific and literary academies. He promoted Brazil as offering the best of European culture with a more pleasant climate, exotic plants and animals, and enough land that a person need never see his neighbor. One of the tens of thousands of immigrants lured by this image was Francois Honoré Dumont, a Parisian jeweler, who moved with his wife to Brazil. He was Santos-Dumont's paternal grandfather.

Brazil's new residents and its Portuguese population benefited from the royal presence in Rio de Janeiro, but disciples of Tiradentes continued to incite rebellion among native Brazilians who felt themselves to be second-class citizens. There were sporadic acts of insurrection, but none really threatened Dom João's rule. The king's most serious challenge was in Portugal itself. Since Napoleon's defeat at Waterloo on June 18, 1815, there had been a power vacuum in Lisbon. In April 1821, Dom João, fearing that someone in Portugal might try to usurp his throne, sailed home with five thousand loyal supporters. He left behind his son Pedro as prince regent, a decision he would regret. Pedro was alarmed by the growing independence movement in Brazil itself, and he made the pragmatic decision to sever ties with both his father and Portugal and declare Brazil an independent entity. On December 1, 1821, the disloyal son, only twenty-four years of age, was crowned the first emperor of Brazil, making the country the only constitutional monarchy in Latin America, among republics cast off from the Spanish Empire. It would remain that way until 1899.

Dom Pedro I's reign lasted a decade. He was more imperial than his father and showed little interest in working with the legislature empowered by Brazil's new constitution. In 1831, faced with congressmen conspiring against him, he abdicated his throne, fled the country, and left behind as the sovereign his five-year-old son, also named Pedro. The constitution provided that "little Pedro" could not be crowned until he was eighteen, but the legislature swore him in four years earlier because the absence of an official emperor was fueling political instability.

Alberto Santos-Dumont was born during Pedro II's reign, on July 20, 1873, in a remote outpost of Minas Gerais, the home state of Tiradentes. Alberto's parents, Henrique Dumont and Francisca de Paula Santos, were first-generation Brazilians who lived in a district of the state named João Aires. The town they resided in was Cabangu, but to call it a town exaggerated its size. At first Cabangu consisted of only their house. Henrique was an engineer, and he had won a contract to build a stretch of the Dom Pedro II railroad through this remote part of Minas Gerais. The railroad was one of the vast public-works projects that the emperor had planned, and so it was an honor for Henrique to receive the commission. The drawback was the necessity of living in isolation.

By the time Alberto was six, the railway work was over, and his father, tapping into his wife's inheritance, had moved the family south to fertile lands outside São Paulo and entered the coffee business. It was not an easy move. The land had to be cleared, five million coffee trees planted, elaborate facilities built to store, dry, and process the beans, and living quarters erected for the workers and foremen. The plantation was so large that Henrique built sixty miles of railway to span it and purchased seven locomotives. The work paid off. Henrique, nicknamed the coffee king by the press, soon had one of the country's largest farms. With his newfound wealth, he could afford to import European tutors for his children and to send Alberto, when he was older, to private schools in São Paulo and Ouro Prêto.

"Inhabitants of Europe comically picture these plantations to themselves as primitive stations of the boundless pampas, as innocent of the cart and wheelbarrow as of the electric light and telephone," Santos-Dumont wrote when he was an adult. "There are such stations far in the interior. I have been through them . . . but they are not the coffee plantations of São Paulo. I can hardly imagine a more stimulating environment for a boy dreaming over mechanical inventions." At the age of seven he drove the broad-wheeled "locomobiles," the steam traction engines that carried the red coffee berries from the fields to his father's railway. Five years later he sweet-talked a foreman into letting him fire up a huge Baldwin locomotive and haul a trainload of berries to the processing plant.

Of Henrique's eight children, it was Alberto, the sixth child and youngest of three sons, who took the most interest in the mechanics of coffee production. He knew every step in the long process. "I think it is not generally understood how scientifically a Brazilian coffee plantation may be operated," he recalled, with the berries untouched by human hands from the moment they entered the train cars to the time their by-product was loaded onto transatlantic ships. In a memoir written in 1904, called *My Air-Ships*, Santos-Dumont lovingly described how his family processed the coffee. The first step was to transfer the berries to giant tanks

where the water is continually renewed and agitated. Mud that has clung to the berries from the rains, and little stones, which have been mixed with them in the loading of the cars, go to the bottom, while the berries and the little sticks and bits of leaves float on the surface and are carried from the tank by means of a trough, whose bottom is pierced with innumerable little holes. Through these holes falls some of the water with the berries, while the little sticks and pieces of leaves float on.

The fallen coffee berries are now clean. They are still the color and size of cherries. The red exterior is a hard pod, or *polpa*. Inside of each pod are two beans, each of which is covered with a skin of its own. The water which has fallen with the berries carries them on to the machine called the *despolpador,* which breaks the outside pod and frees the beans. Long tubes, called "driers," now receive the beans, still wet and with their skins on them. In these driers the beans are continually agitated in hot air.

Coffee is very delicate. It must be handled carefully. Therefore the dried beans are lifted by the cups of an endless-chain elevator to a height whence they slide down an inclined trough to . . . the coffee machine house.

The beans were then carried by a second elevator to the processing plant. The first machine they encountered was a ventilator, a series of vibrating sieves that let the coffee beans slip through but trapped any remaining sticks, leaves, and pebbles, impurities that would break the next machine.

Another endless-chain elevator carries the beans to a height whence they fall through an inclined trough into the *descascador*, or "skinner." It is a highly delicate machine; if the spaces between are a trifle too big, the coffee passes without being skinned, while if they are too small, they break the beans.

Another elevator carries the skinned beans with their skins to another ventilator, in which the skins are blown away.

Still another elevator takes the now clean beans up and throws them into the "separator," a great copper tube two yards in diameter and about seven yards long, resting at a slight incline. Through the separator tube the coffee slides. As it is pierced at first with little holes, the smaller beans fall through them. Further along it is pierced with larger holes, and through these the medium-sized beans fall; and further still along are yet larger holes for the large, round beans called "mocha." Each grade falls into the hopper, beneath which are stationed weighing scales and men with coffee sacks. As the sacks fill up to the required weight, they are replaced by empty ones; and the tied and labeled sacks are shipped to Europe.

As a boy, Santos-Dumont spent whole days watching the machines and teaching himself to fix them. They were always breaking down.

In particular, the moving sieves were continually getting out of order. While they were not heavy, they moved back and forth horizontally at great speed and took an enormous amount of motive power. The belts were always being changed, and I remember the fruitless efforts of all of us to remedy the mechanical defects of the device.

Now, is it not curious that these troublesome shifting sieves were the only machines at the coffee works that were not rotary? They were not rotary, and they were bad. I think this put me as a boy, against all *agitating* devices in mechanics, and in favor of the more easily handled and more serviceable rotary movement. . . .

This was a prejudice that would serve him well when he built flying machines as an adult.

Alberto was also the handyman around the house. His mother's sewing machine often jammed, and she expected him to stop whatever he was doing and repair it. When arms or legs fell off his sisters' dolls, he was the one who reattached them. When the wheels on his brothers' bicycles started to wobble, he was the one who realigned them.

Alberto was a loner and a dreamer, preferring the company of plantation machinery to a meal with his family. The atmosphere in Alberto's house was often tense. His mother was deeply religious and superstitious, and his father, a man of reason and science, openly mocked her beliefs at family dinners. Although Henrique was pleased by his youngest son's fascination with technology, he did not understand why Alberto had no interest in hunting, roughhousing, and the other manly activities that his brothers liked. Alberto never joined the men on all-day horseback expeditions and picnics to the far reaches of the land.

At night he stayed up late reading. His father, who had received his engineering training in Paris at the Ecole Centrale des Arts et Métiers, had stacks of books lying around the house, in French, English, and Portuguese. Alberto paged through most of them, even the technical manuals. His favorites were science fiction. He loved Jules Verne's vision of a sky populated with flying machines and had read all his novels by the age of ten. He learned from his father's engineering texts that the hot-air balloon had been invented in 1783, by Joseph and Etienne Montgolfier, papermakers in Annonay, France, a town in the Rhône valley forty miles from Lyons. The brothers had constructed a large pear-shaped envelope, from paper or silk, with an aperture at the base so that it could be inflated with smoke from burning straw. One account said that the inspiration had come from Joseph's aimlessly tossing the conical paper wrapping of a sugarloaf into the fireplace and then being surprised to see it rise up the chimney without igniting. Another story attributed it to his watching his wife's camisole levitate after she hung it in front of the hearth to dry.

The fact that "millions of people" over the ages must have observed similar phenomena, one commentator noted, "and had *not* derived anything practical therefrom only enhances the glory of those who in such well-worn tracks did make a discovery." The earliest suggestion of aerostation, as ballooning was called, predated the Montgolfiers by two thousand years but was probably not authentic. In *Noctes Atticae* ("Attic Nights"), the Roman writer Aulus Gellius described a flying dove constructed by Archytas of Tarentum, a Pythagorean mathematician in the fourth century B.C. It was a "model of a dove or pigeon formed in wood and so contrived as by a certain mechanical art and power to fly: so nicely was it balanced and put in motion by hidden and enclosed air." Although the "hidden and enclosed air" suggested an anticipation of the hot-air balloon, it was doubtful that a hollow wooden bird would have been light enough to ascend. It was more likely that the dove's apparent flying was a mechanical trick accomplished by invisible wires.

The physical basis of aerostation was as simple as the Montgolfiers' solution of imprisoning hot air in a bag: The balloon floated because it weighed less than the equivalent volume of air, just as a seafaring ship floated because it weighed less than the equivalent volume of water. But the analogy between ship and balloon worked only if one accepted the idea that the atmosphere weighed something, and that was not known until Galileo's time, when Evangelista Torricelli, the inventor of the barometer, demonstrated that the atmosphere had a measurable weight that decreased with elevation. Another seventeenth-century investigator, Otto von Guericke in Magdeburg, Germany, invented a vacuum pump for creating the "rarefied air" found at very high elevations. In 1670, Francesco de Lana-Terzi, an Italian Jesuit priest, conceived of a man-carrying vessel supported by four huge hollow-copper spheres devoid of air. Because the evacuated spheres would be lighter than the air they displaced, he expected the vessel to rise through the atmosphere like an air bubble ascending through water. The mathematically sophisticated priest calculated that the spheres had to be twenty-five feet in diameter and $\frac{1}{225}$ of an inch in thickness. When his physicist friends warned him that spheres this thin would collapse when the air was withdrawn from them, he responded—according to engineering historian L. T. C. Rolt—"that his was only a theoretical exercise, arguing that since God had not intended man to fly, any serious practical attempt to flout His designs must be impious and fraught with peril for the human race. One suspects that the Jesuit fathers may have had a serious talk with their scientifically minded son and that he made this disclaimer because he could smell faggots burning."

But other clerics continued the armchair exercise. In 1755, Joseph Galien, a Dominican friar and theologian at the papal university in Avignon, proposed collecting rarefied air from the upper reaches of the atmosphere and enclosing it in a mile-long vessel that would be capable of lifting fifty-four times the weight carried by Noah's ark. Galien never explained how he planned to reach the upper atmosphere in the first place, and his supervisor at the divinity school implored him to take a long respite from clerical duties and, on his return, to restrict his speculation to theology not technology.

Such chimerical schemes for ballooning were abandoned once the Montgolfiers showed how little there really was to it. On June 5, 1783, the two brothers demonstrated a thirty-foot-diameter unmanned balloon in the public square in Annonay. It required eight men to hold down the twenty-thousand-cubic-foot balloon, whose envelope consisted of pieces of silk lined with paper fastened together by buttons and button-holes. When the Montgolfiers gave the signal, the men released the giant gasbag and it climbed six thousand feet. After ten minutes, it came down in a field a mile and a half away.

News of the accomplishment reached the Paris Academy of Sciences, whose members had been actively experimenting with the construction of a lighter-than-air balloon but had so far failed to get anything off the ground. The Parisian scientists, not wanting to be upstaged by unschooled papermakers, accelerated their efforts. The physicist-engineer Jacques Alexandre César Charles, assisted by two craftsmen, the brothers Ainé and Cadet Robert, substituted hydrogen gas for the burning-straw fumes, and on August 23, 1783, in the place des Victoires began inflating a twelve-foot-diameter silk balloon. The hydrogen was obtained by pouring five hundred pounds of sulfuric acid over one thousand pounds of iron filings. Charles had not counted on the chemical reaction to produce as much heat as it did, and the balloon fabric had to be repeatedly doused with cold water to keep it from singeing. The water vapor trapped in the balloon condensed and weighed it down.

The inflation took three days, and, as word spread of the spectacle, a crowd gathered, choking the neighboring streets. To ease the congestion, Charles ordered the balloon moved in the stealth of night, escorted by armed guards, to the more expansive Champ de Mars, at the foot of what is now the Eiffel Tower. Barthélemy Faujas de Saint-Fond witnessed the move:

No more wonderful scene could be imagined than the Balloon being thus conveyed, preceded by lighted torches, surrounded by a "cortege" and escorted by a detachment of foot and horse guards; the nocturnal march, the form and capacity of the body, carried with so much precaution; the silence that reigned, the unseasonable hour, all tended to give a singularity and mystery truly imposing to all those who were acquainted with the cause. The cab-drivers on the road were so astonished that they were impelled to stop their carriages, and to kneel humbly, hat in hand, whilst the procession was passing.

At 5:00 P.M. on August 27, Charles's assistants triumphantly released the balloon and it rose rapidly to a height of three thousand feet. After forty-five minutes, it descended in a field in the village of Gonesse, fifteen miles from Paris.

Unlike the hot-air balloon, which could have been made at any time in recorded history, the hydrogen balloon could not have been invented much earlier than it was because the gas, initially called phlogiston, or "inflammable air," was discovered only in 1766, by the English scientist Henry Cavendish. On learning that "inflammable air" was nine times lighter than ordinary air, Joseph Black in Edinburgh filled a small, thin bag with the new gas and watched it rise to the ceiling of his laboratory. He had difficulty, though, in scaling up the experiment. The problem was that the materials he tried for bags were either too heavy or too porous. At a large public lecture, he used the allantois of a calf as the gasbag, but was humiliated by its failure to ascend and gave up ballooning entirely. In 1782, Tiberius Cavallo, a fellow of London's Royal Society, "found that bladders, even when carefully scraped, are too heavy, and that China paper is permeable to the gas." Charles succeeded because he had the idea of making the silk impermeable but still lightweight by varnishing it with a solution of elastic gum.

The Montgolfiers made the next move in the race to advance aerostation. On September 19, 1783, they repeated the Annonay experiment at Versailles for the benefit of Louis XVI, Marie Antoinette, and their court. According to one observer, the papermakers "had caused all the old shoes that could be collected to be brought here, and threw them into the damp straw that was burning, together with pieces of decomposed meat; for these are the substances which supply their gas. The King and the Queen came up to examine the machine, but the

noxious smell thus produced obliged them to retreat at once."
French scientists found the demonstration particularly insulting
because the two brothers had beaten them to the balloon's
invention while harboring incorrect notions about the cause of
the ascension. The Montgolfiers attributed the "lifting power"
to the lighter-than-air smoke generated from their patented
combination of fetid meat and dirty shoes. In fact, the smoke
particles were heavier than air and actually worked to coun-
teract the balloon's rise. The lift came not from the imprisoned
smoke but the captured hot air, which was lighter than the
cooler ambient air. Most of the observers did not care why
the spectacular blue-and-gold balloon was aloft—they just marveled at the fact that it was. And the world's first aerial travelers, a sheep, a rooster, and a duck, were suspended in a cage
below the balloon. The animals emerged unscathed from their
two-mile trip to the forest of Vaucresson, except for the
rooster, whose right wing had suffered a nasty kick from the
sheep.

Charles and the Montgolfiers independently told the king
that on the next ascension they themselves would be the passengers, but his majesty forbade such valuable subjects from
risking their lives. Instead he offered prisoners as the first pilots, proposing to set them free if they survived. But Charles
ultimately convinced him that the first person aloft should be
a man of science who could describe the voyage if he were
fortunate enough to make it back. The honor went to Francis
Pilâtre de Rozier, a distinguished member of the Academy of
Sciences who was the superintendent of the king's natural-
history collection. On October 15, 1783, he ascended in a
captive balloon (one tethered to the ground), the hot air
replenished by the burning of straw and wood in an iron basket hung below the balloon. Having found it easy to stoke the

fire when he was in the air, Pilâtre de Rozier and a companion, the Marquis d'Arlandes, went up in a free balloon for the first time on November 21. Ascending from the Bois de Boulogne at 1:54 P.M., they reached an elevation of five hundred to one thousand feet and, after twenty-five minutes, descended beyond the Paris city limits, some nine thousand yards from where they had started. Ten days later, Charles and Ainé Robert had the honor of being the first people to ascend in a hydrogen balloon, in a two-hour journey that began in the Tuileries and ended twenty-seven miles away in the town of Nesle.

Within a few months of Charles's trip, the skies of Paris were populated with both hydrogen balloons, known as *charlières*, and *montgolfières* (hot-air balloons). *Charlières* were safer because they did not require an open flame, but *montgolfières* were more practical because hydrogen was expensive and scarce. "Balloonomania," as historian Lee Kennett called it, was sweeping France: "The decade of the 1780s was in many ways a frivolous and jaded age, and it took the new 'aerostatic machines' to its heart. Ascensions became as fashionable as costume balls, and so numerous that the Paris city authorities had to issue an ordinance governing their use—the world's first air traffic regulations. The distinctive form of the balloon lent itself to objects as diverse as chair backs and snuff boxes."

IN 1883, Alberto Santos-Dumont, age ten, had not yet seen a balloon, but he duplicated the Mongolfiers' invention in miniature. Working from illustrations in books, he made handheld balloons out of tissue paper and filled them with hot air from the stove flame. At holiday celebrations he demonstrated the gasbags to the field hands. Even his parents, who did not approve of his incendiary experiments, could not conceal their amazement when the *montgolfières* soared higher than the house. He also made a toy wooden plane whose propeller, or "air screw" as it was called in those days, was powered by a wound-up rubber string.

From reading Verne, Alberto was convinced that people had already gone beyond the hot-air balloon and flown airships, also known as dirigibles (steerable powered balloons). His family and childhood friends tried to disabuse him of the notion. He used to play a game with the other children called Pigeon flies! One boy was chosen as the leader, and he would shout, "Pigeon flies! Hen flies! Crow flies! Bee flies!" and so on. "At each call we were supposed to raise our fingers," Santos-Dumont wrote many years later. "Sometimes, however, he would call out: 'Dog flies! Fox flies!' or some other like impossibility to catch us. If anyone raised a finger, he was made to pay a forfeit. Now my playmates never failed to wink and smile mockingly at me when one of them called 'Man flies!' for at the word I would always lift my finger very high, as a sign of absolute conviction; and I refused with energy to pay the forfeit. The more they laughed at me, the happier I was, hoping that someday the laugh would be on my side."

It was not until Alberto was fifteen that he actually saw a manned balloon. At a fair in São Paulo, in 1888, he watched a performer ascend in a nearly spherical gasbag and descend by parachute. Alberto's imagination took off:

In the long, sun-bathed Brazilian afternoons, when the hum of insects, punctuated by the far-off cry of some bird, lulled me, I would lie in the shade of the veranda and gaze into the fair sky of Brazil, where the birds fly so high and soar with such ease on their great outstretched wings, where the clouds mount so gaily in the pure light of day, and you have only to fall in love with space and freedom. So, musing on the exploration of the aerial ocean, I, too, devised airships and flying machines in my imagination.

These imaginings I kept to myself. In those days, in Brazil, to talk of inventing a flying machine, or dirigible balloon, would have been to stamp one's self as unbalanced and visionary. Spherical balloonists were looked on as daring professionals not differing greatly from acrobats; and for the son of a planter to dream of emulating them would have been almost a social sin.

Santos-Dumont's parents were politically conservative. They supported the emperor, whose railroad Henrique had eagerly constructed. But they could not keep their curious son from being exposed to all sorts of ideologies that they found distasteful. When Alberto was in the coffee-processing plant, even though he generally kept to himself, he would overhear conversations. Sometimes the workers talked about the democratic movement and spoke with passion of the patriot Tiradentes. The revolutionary dentist had become the hero of ordinary Brazilians, and his life was being turned into myth, as would Santos-Dumont's years later. Tiradentes was depicted in numerous paintings as a bearded Christ-like figure, although in reality he was clean-shaven and short-haired. The day of his execution, April 21, became a national holiday, which is still celebrated today. Young Alberto had little interest in politics, and obviously no desire to be drawn and quartered, but he was attracted to the immortality that Tiradentes had achieved. He decided then that he wanted to do something with his life that would stir the hearts of men and women— an extraordinary aspiration for an adolescent to have. He had no idea what profession he would take up—it may not even have crossed his mind that one could become an aeronaut or an inventor. But he knew that whatever he did, it should have a profound impact on the people around him. Certainly no other aeronautical pioneer had such grand ambitions a decade before taking to the air.

[CHAPTER 2]

"A MOST DANGEROUS

PLACE FOR A BOY"

PARIS, 1891

•

SANTOS-DUMONT'S INSULAR world expanded when he was eighteen. His sixty-year-old father, who still lorded over both his family and the plantation, was thrown from his horse and suffered a severe concussion and partial paralysis. When he did not fully recover, Henrique abruptly sold the coffee business for $6 million and headed to Europe in search of medical treatment with his wife and Alberto in tow. The threesome took a steamer to Lisbon. After a brief respite in Oporto, where two of Alberto's sisters had taken up residence with their Portuguese husbands, two brothers by the name of Villares (and a third sister back in Brazil was married to yet another Villares brother), they boarded a train for Paris. Henrique had faith that the city's doctors would cure him. After all, it was the place where Louis Pasteur was performing medical miracles, saving children from rabid canines by vaccinating them.

From the moment in 1891 when Santos-Dumont disembarked at the Gare d'Orléans, he fell in love with the city. "All good Americans are said to go to Paris when they die," he wrote. For a teenager who loved inventions, fin de siècle Paris represented "everything that is powerful and progressive." He lost no time immersing himself in the city's technological wonders. On his first day, he visited the two-year-old Eiffel Tower, which at 986 feet stood almost twice as tall as any other man-made structure in the world. Although the massive iron latticework was illuminated by conventional gas lighting, the elevators that carried sightseers and meteorologists to the observation deck were powered by that exciting new form of energy –electricity. Alberto rode the elevators for half a day, and then he sat on the bank of the Seine and admired the tower's sky-high curves.

Henrique shared his delight. When he was trained as an engineer four decades earlier, the profession did not enjoy the exalted reputation it now had in France and England. The construction of strong but graceful bridges to extend railway systems across the rivers and gorges of Europe had elevated the status of the engineer. "If we want any work done of an unusual character and send for an architect, he hesitates, debates, trifles," Prince Albert of Great Britain observed. "Send for an engineer, and he *does* it." Gustave Eiffel was one of the master bridge builders, and he won the commission to construct the monumental tower for the Paris Exposition of 1889, a world's fair that celebrated the centennial of the French Revolution and the benefits of nineteenth-century industrialization. On both sides of the Atlantic, there had been talk of building a thousand-foot tower, but the will to do it was strongest in France. Paris wanted to prove to itself and the world that it had recovered fully from both the 1871 Franco-Prussian War, in which the Germans had annexed the eastern provinces of Alsace and Lorraine, and the subsequent Paris Commune, in which twenty thousand Frenchmen had been slaughtered by their fellow Parisians and whole sections of the city leveled.

Exposition planners blessed Eiffel's blueprints as soon as they saw them, but a few vocal writers and painters protested the idea of a "dizzily ridiculous tower dominating Paris like a black and gigantic factory chimney" with no escape from "the odious shadow of the odious column of bolted metal." But once the tower was actually constructed, most of the indignant aesthetes came around to liking it, with the notable exception of the writer Guy de Maupassant, who, it was said, dined regularly "at the restaurant on the second platform because that was the only place in the city where he could be certain not to see the tower." In 1891, Parisians were still in the midst of their honeymoon with the ten-thousand-ton, wrought-iron giant. Henrique and Alberto watched fashionable young ladies climb the 1,671 stairs in fanciful dresses from rue Auber known as the *Eiffel ascensionniste,* which boasted a series of nested collars to protect "the adventurous wearer against cooler temperatures at high altitude."

Alberto Santos-Dumont also marveled at the novel vehicles he saw. The first mass-produced bicycles rolled quietly along the streets, with rubber tires in place of the clacking wooden wheels with which he was familiar. The bicycle gave middle-class Parisians a form of mobility that few Brazilians could afford, and contributed to a sexual revolution when women, demanding the same freedom of movement as men, insisted on their own bikes and, to ride them, donned pants—culottes—for the first time. (A popular advertisement of the time depicted a grinning bride speeding away on a bicycle after abandoning her beau at the altar.) The first few motorcars, totally unknown in Rio, clamored down the boulevards at speeds of less than ten miles an hour—and provoked the same artists who had objected to the Eiffel Tower to sniff that "the harsh smell of gasoline obliterates the noble smell of horse manure." On the street corners were *théâtrophones,* special pay phones by which Parisians could listen to live opera, chamber music, plays, even political meetings.

Despite the conspicuous new technologies, the typical apartment house, except in the so-called American quarter on the right bank, lacked certain conveniences that were already common in New York and Chicago (but not yet in Rio or São Paulo). "Elevators are the exception rather than the rule, candles are more in evidence than incandescent lamps . . . and such a thing as a well-equipped bathroom is practically nonexistent," observed New Yorker Burton Holmes, a contemporary of Santos-Dumont and one of the world's first photojournalists. Holmes was particularly vexed by the difficulty of taking a hot bath:

> "Un bain, Monsieur? Mais parfaitement! I will make the bath to come at five o'clock this afternoon," said the obliging *concierge* when I expressed a desire for total immersion. "But I want the bath now, this morning, before breakfast," I insisted. "Impossible, Monsieur, it requires time to prepare and to bring, but it will be superb—your bath—the last gentleman who took one a month ago enjoyed his very much. You will see, Monsieur, that when one orders a bath in Paris, one gets a beautiful bath—it will be here at four o'clock." At four, a man, or rather a pair of legs, came staggering up my stairs—five flights, by the way—with a full-sized zinc bath tub, inverted and concealing the head and shoulders and half the body of the miserable owner of those legs. The tub was planted in the middle of my room: a white linen lining was adjusted; sundry towels and a big bathing sheet, to wrap myself in after the ordeal, were os-

tentatiously produced. Then came the all-important operation of filling the tub. Two pails, three servants, and
countless trips down to the hydrant, several floors below,
at last did the trick: the tub was full of ice-cold water. "But
I ordered a hot bath." "Patience, Monsieur, behold here
is the hot water!" Whereupon the bath man opens a tall
zinc cylinder that looks like a fire extinguisher and pours
about two gallons of hot water into that white-lined tub—
result a tepid bath—expense sixty cents—time expended
two hours, for the tub had to be emptied by dipping out
the water and carrying it away, pail after pail. Then the
proud owner of the outfit slung his pails on his arms, put
his tub on his head like a hat, and began the perilous
descent of my five flights of stairs.

In private homes the telephone was as scarce as hot water.
"Polite society proved relatively slow to accept the phone,"
historian Eugen Weber noted, and even "President Grévy took
a lot of persuading before he allowed one to be installed in
the Elysée Palace." The upper class regarded the phone as
interfering with the sacred privacy of their living space. Rare
was a Parisian like the Comtesse Greffulhe who appreciated
"the magic, supernatural life" that the phone provided: "It's
odd for a woman to lie in bed," she explained, "and talk to a
gentleman who may be in his. And you know, if the husband
should walk in, one just throws the thingummy under the bed,
and he does not know a thing." As late as 1900, "there were
only 30,000 telephones in France," Weber observed, when
New York City's hotels had more than 20,000 among them.

And yet, with the exception of a few grumbling aesthetes,
Parisians, even more so than New Yorkers, had an abiding
faith in the inherent goodness of technology. When New York

State introduced the electric chair in 1899, Weber said, the power companies objected, fearing that if people knew that electricity could kill, they would not want it in their homes or offices. The French, on the other hand, laughed off the possibility of a deadly electric chair; they could not imagine that such a wondrous new source of power could be destructive.

Santos-Dumont felt at home among Paris's technophiles. The city had everything going for it, he thought, except that the sky was astonishingly deficient in airships. He expected it to be peppered with real-life versions of Verne's flying machines. This after all was the country where the Montgolfier brothers had sent up the first hot-air balloon a century before. Moreover, as Santos-Dumont knew, in 1852 a Frenchman named Henri Giffard had chugged along at half a mile an hour in the world's first powered balloon by hanging a five-horsepower steam engine and propeller from a 144-foot-long cigar-shaped gasbag. In 1883 two brothers, Gaston and Albert Tissander, had substituted an electric motor and boosted the speed to three miles an hour. As part of the French military's balloon program, Colonel Charles Renard and Lieutenant Arthur Krebs had more success with an electric engine in 1884, setting a speed record of 14.5 miles an hour. Santos-Dumont could not understand why in the ensuing seven years the airship had not evolved into an everyday conveyance. Indeed, it had devolved: There were no airships at all in 1891.

The powerless gasbags that did inhabit the sky were generally tethered, anchored by long cords that kept them from drifting away. Most of these balloons were not operated by inventors or men of science but by street performers. One woman of particular renown sat at a piano suspended from a balloon and played Wagner five hundred feet above the ground. Another showman regularly sent up roosters, turtles, and mice and prided himself that they were none the worse for it. In Paris there were also a few shameless hucksters who charged exorbitant fees for rides in untethered balloons. They could control the elevation more or less by throwing out ballast or letting out gas, but they had little influence on where the wind might sweep them.

In earlier times, clerics had railed against men who tried to fly, warning them that they were flirting with disaster by encroaching on the realm of the angels. In 1709, the Brazilian aeronaut Laurenco de Gusmao, known as the flying priest, was put to death as a sorcerer by the Inquisition. Even in enlightened fin de siècle France, this view of flying as black magic persisted among the lower classes. Santos-Dumont had heard the story of an errant balloon that was carried by an unpredictable wind from Paris to a nearby town, where it precipitously crashed. As the unlucky paying customer climbed nervously out of the basket, peasants attacked the limp gasbag, beating it ferociously with sticks and denouncing it as devil's work. To prevent future incidents that might end even more violently, the government distributed a pamphlet in the countryside explaining that balloons were not vessels of the dark forces. Santos-Dumont thought there must be a better way. He decided it was his mission to design a steerable balloon that could fight the wind so that no one would be swept inadvertently onto a stranger's land.

The first step, he decided, was to go up in one of the existing balloons. On a day when his parents were occupied getting medical advice about his father's condition, Santos-Dumont looked up *balloonist* in the city directory and visited the first one listed.

"You want to make an ascent?" the man asked gravely.

"Hum, Hum! Are you sure you have the courage? A balloon ascent is no small thing, and you seem too young."

Santos-Dumont assured him of his purpose and his resolve, and the aeronaut consented to take him up for at most two hours provided that the day was sunny and the skies calm. "My honorarium will be twelve hundred francs [two hundred and forty dollars]," he added, "and you must sign a contract to hold yourself responsible for all damage we may do to your own life and limbs, and to mine, to the property of third parties, and to the balloon itself and its accessories. Furthermore, you must agree to pay our railway fares and transportation for the balloon and its basket back to Paris from the point at which we come to the ground."

Santos-Dumont asked for time to think it over. "To a youth eighteen years of age," he wrote in his memoir, "twelve hundred francs was a large sum. How could I justify it to my parents? Then I reflected: 'If I risk twelve hundred francs for an afternoon's pleasure, I shall find it either good or bad. If it is bad, the money will be lost. If it is good, I shall want to repeat it and I shall not have the means.' This decided me. Regretfully, I gave up ballooning and took refuge in automobiling"—an interest that was piqued when he accompanied his father to the Palais des Machines, a building that, like the Eiffel Tower, was constructed as part of the Paris Exposition of 1889. During the exposition, the cavernous building, an iron-and-glass cathedral to technology, housed thousands of exhibits from all over the world, from mining equipment and steam-powered looms to the first gas-powered automobile, patented by Karl Benz, and Thomas Edison's display of phonographs and electric lights, operated by the inventor himself. Even though the exposition had officially ended months before

Henrique and Alberto's visit, the Palais des Machines continued to house new technologies. At one point, Henrique realized that he had lost his son. He meandered slowly back through the hall in his wheelchair and found Alberto mesmerized by a working internal combustion engine, entranced that a machine much smaller than a steam engine could be so powerful. "I stood there as if I had been nailed down by Fate," Santos-Dumont recalled. "I was completely fascinated. I told my father how surprised I was at seeing that motor work, and he replied: 'That is enough for today.' "

Alberto subsequently visited the Peugeot workshop in Valentigny. Although he had reservations about spending his father's hard-earned money on a balloon flight, he had none about spending it on a 3.5-horsepower four-wheeler. Peugeot manufactured only two cars in 1891—the steering and brakes barely worked—and the eighteen-year-old Brazilian was now the proud owner of one of them. In a few months, when his father came to the realization that Parisian medicine could not restore his health, Alberto sailed with him back to Brazil. Alberto brought along the Peugeot roadster, and when he powered it up in São Paulo, he reportedly had the distinction of being the first person ever to drive a car in South America.

Henrique knew that he was dying, so he had a long talk with Alberto about his future. He had seen how happy his youngest son had been in the City of Light, and, to his wife's chagrin, urged him to return to Paris by himself, despite cryptically warning him that the city was "a most dangerous place for a boy." He told Alberto that he did not have to worry about earning a living and advanced him his inheritance of a half million dollars. He sent him off with the challenge "let's see if you make a man of yourself"—strong words that reflected his concern that his son had never shown the slightest interest in the opposite sex. Alberto returned to Paris in the summer of 1892, and his father died in August.

Santos-Dumont's first order of business in Paris was to look up other balloonists in the city directory. But "like the first," he wrote, "all wanted extravagant sums to take me up with them on the most trivial kind of ascent. All took the same attitude. They made a danger and a difficulty of ballooning, enlarging on its risks to life and property. Even in presence of the great prices they proposed to charge me, they did not encourage me to close with them. Obviously they were determined to keep ballooning to themselves as a professional mystery. Therefore, I bought a new automobile."

He also attended to his education. He and his father had investigated the colleges in Paris, but in the end his father, knowing that Alberto might rebel against a structured curriculum, suggested that he hire a private tutor instead. That was fine with Alberto, who had recurrent nightmares about being called upon to answer a question in a crowded classroom. In 1892, he employed a former college professor named Garcia and the two of them designed an intense program of study weighted toward the "practical sciences"—physics, chemistry, and mechanical and electrical engineering. The home-study plan appealed to the recluse and bookworm in Santos-Dumont, and for the next five years he buried himself in his textbooks. Occasionally he visited cousins in England, where he would slip into the back of lecture halls at the University of Bristol and listen to the professors; because he was not an official student, there was little danger of his being called on.

For relaxation in those studious years, Santos-Dumont drove his cars. (According to Brazilian biographies, he owned more cars in 1892 than anyone else in Paris—but the truth of this assertion, and just how many cars he owned, cannot be confirmed.) He would pass the time motoring up and down the wide boulevards, but the first internal combustion engines were so unreliable that they often broke down, stalling the predominant horse traffic. His Peugeot was such a novelty that even when it was running properly, it snarled traffic as pedestrians rushed into the street to get a better view. The police warned him to keep the car moving, and once he was fined—in what might have been the city's first traffic violation—for causing a disturbance near the opera house. The "disturbance" was actually a convivial affair, an impromptu street party of passersby tickled by the sight of his vehicle.

Fin de siècle Parisians were great revelers. According to Eugen Weber, they turned even unpleasant experiences, such as being vaccinated against smallpox, into festive occasions. Inoculations were "done at parties, as if one were going to the theater," the society columns reported. "One organizes an intimate luncheon; the doctor arrives at dessert, the vaccine in his pocket." In his student days, Santos-Dumont was not much for revelry or heavy drinking, but he occasionally indulged in Parisian nightlife, that inebriated swirl of intelligent discourse and decadence. The cafés were places for learned conversation not just about art and literature but scientific and technological developments like the discovery of X rays and the construction of the Paris metro. People who fashioned themselves intellectuals stayed up all night talking, pausing to inject themselves with morphine in gold-plated syringes, sip Vin Coca Mariani (a wine permeated with cocaine), or snack on strawberries soaked in ether. End-of-the-century Paris was forgiving to someone like Santos-Dumont who was not sure about his sexuality. The avant-garde café crowd promoted erotic experimentation, and homosexuality became so fashionable that every hipster had to try it. "All remarkable women do it," the wife of a banker wrote, "but it's very difficult. One has to take lessons."

In 1897, Santos-Dumont returned home to Brazil and reflected on his five years in Paris. With Garcia's guidance, he had mastered the sciences. He was thankful for that, and yet there were things he wished he had done. "I regretted bitterly that I had not persevered in my attempt to make a balloon ascent," he wrote. "At that distance, far from ballooning possibilities, even the high prices demanded by the aeronauts seemed to me of secondary importance."

Before heading back to Paris, he visited a bookstore in Rio and purchased a copy of *Andrée's Balloon Expedition, In Search of the North Pole*. The book, written by the Paris balloon makers Henri Lachambre and Alexis Machuron, proved to be a great diversion on the long steamship passage. Lachambre and Machuron had built a huge balloon called the *Eagle* for the young Swedish scientist Salomon August Andrée, who had been planning for more than a decade to make the first balloon expedition to the North Pole. Andrée finally got the opportunity on July 11, 1897, when he ascended from Dane's Island, off the north coast of Norway near Spitsbergen, for a 2,300-mile journey that he hoped to complete in six days. Accompanying him were two companions, three dozen carrier pigeons, a boat, a stove, sleds, tents, sundry scientific instruments, cameras, and enough food and birdseed to last four months. Although the *Eagle* was not powered, Andrée had cleverly equipped it with large sails so that he could steer it on a course that could deviate by as much as thirty degrees from the wind.

Lachambre and Machuron had published their book within days of Andrée's ascent, before it was evident what had happened to him. They reported that one carrier pigeon had delivered an encouraging message: "13th July, 12:30 P.M., 82.2° N. Lat., 15.5° E. Long. Good progress toward the north. All goes well on board. This message is the third by pigeon. Andrée." After Santos-Dumont disembarked in France, he learned that only one other pigeon had made it back. The Andrée expedition was the talk of Paris cafés. The prevailing sentiment was that he would not return, and indeed that turned out to be the case. Three decades would pass before a hunting party discovered Andrée's body and diary on White Island, a deserted expanse of pack ice only 150 miles from the *Eagle*'s starting point. The sails had apparently failed, and Andrée could not steer the balloon out of a fierce snowstorm that finally forced it down. He described in the diary how he and his companions had survived on lichen and seal blubber for three months. Then the journal entries ended. The brutal winter had set in, and the men froze to death in a blizzard.

In his own journal, Santos-Dumont noted how much Andrée's story had affected him: "The reading of the book during the long voyage proved a revelation to me, and I finished by studying it like a textbook. Its description of materials and prices opened my eyes. At last I saw clearly. Andrée's immense balloon—a reproduction of whose photograph on the book cover showed how those that gave it the final varnishing climbed up its sides and over its summit like a mountain—cost only 40,000 francs to construct and equip fully! I determined that, on arriving in Paris, I would cease consulting professional aeronauts and would make the acquaintance of constructors."

Santos-Dumont saw a little of himself in Salomon Andrée. He liked Andrée's adventurous spirit and shared his belief in the unbounded power of technology to end human misery. Andrée had described, in a series of sanguine articles, the likely benefits that the electric light and other new inventions would have on human evolution, liberty, hygiene, athletics, language, architecture, military planning, home life, marriage, and education. Despite his loquaciousness in print, Andrée was a man of few words at public functions, and Santos-Dumont too was tongue-tied at formal affairs.

Both men shunned intimate relationships with women and never married. "In married life, one has to deal with factors which cannot be arranged according to a plan," Andrée wrote. "It is altogether too great a risk to bind oneself into a condition of things where another individual would be fully entitled— and what right would I have to repress this individuality?—*to demand the same place in my life that I myself occupied!* As soon as I feel any heart-leaves sprouting, I hasten to uproot them, for I know that any feeling which I allowed to live would become so strong that I should not *dare* to submit to it."

•

IN THE FALL of 1897, Santos-Dumont sought out the builders of Andrée's balloon, in the hope that the architects of such
a risky and fanciful project as the first flight to the North Pole
would be receptive to his aeronautical interests. Santos-
Dumont visited Lachambre and Machuron in their Parc
d'Aérostation in Vaugirard, a town that had been incorporated into Paris. The two men warmed to him at once. They
did not dismiss him as a feckless dreamer; nor did they demand a large fee or exaggerate the obvious dangers of aerostation. "When I asked M. Lachambre how much it would
cost me to make a short trip in one of his balloons," Santos-
Dumont recalled, "his reply so astonished me that I asked
him to repeat it."

"For a long trip of three or four hours," Lachambre said,
"it will cost you 250 francs, all expenses and return of balloon
by rail included."

"And the damages?" Santos-Dumont asked.

"We shall not do any damage!" he replied, laughing.
Santos-Dumont accepted the deal before Lachambre had a

chance to change his mind. Machuron offered to take him up the next day.

Santos-Dumont did not trust any of his beloved motorized vehicles to get him to the ascension on time, so he traveled by horse-cab, arriving early in Vaugirard so that he could watch the preparations. The deflated balloon lay flat and formless on the grass. On Lachambre's order, the workmen turned on the gas and the balloon slowly swelled into a forty-foot-diameter sphere, holding 26,500 cubic feet of gas. By 11:00 A.M. the preparations were complete. A mild breeze was gently rocking the narrow wicker basket; Machuron stood in one corner, and opposite him was the diminutive Brazilian, impatient and fidgety, clutching a large bag of sand ballast so that the basket would not tip too much in the direction of Machuron, who weighed twice as much as he. "Let go, all!" Machuron yelled. The workmen released the balloon, and Santos-Dumont's first sensation in the air was that the wind had ceased altogether.

"The air seemed motionless around us," recalled Santos-Dumont. "We were off, going at the speed of the air current in which we now lived and moved. Indeed, for us, there was no more wind; and this is the first great fact of spherical ballooning. Infinitely gentle is this unfelt movement forward and upward. The illusion is complete: it seems not to be the balloon that moves, but the earth that sinks down and away." The other surprise was that the horizon appeared elevated: "At the bottom of the abyss which already opened 1500 yards below us, the earth, instead of appearing round like a ball, shows concave like a bowl by a peculiar phenomenon of refraction whose effect is to lift up constantly to the aëronaut's eyes the circle of the horizon." He could still make out people on the ground—they looked like ants, he said (a description that is now a cliché but may have been original to him). He could not hear their voices. The only sounds were the faint barking of dogs and the occasional locomotive whistle.

They climbed much higher. A cloud passed in front of the sun, cooling the gas in the balloon, which started to wrinkle and descend, gently at first and then rapidly. "I was frightened," said Santos-Dumont. "I did not *feel* myself falling, but I could *see* the earth coming swiftly up to us; and I knew what that meant!" The two men jettisoned ballast until they stabilized the balloon at an elevation of ten thousand feet. Santos-Dumont had discovered "the second great fact of spherical ballooning—we are masters of our altitude by the possession of a few kilos of sand!" They were now floating above a layer of clouds. "The sun cast the shadow of the balloon on this screen of dazzling whiteness," he recalled, "while our own profiles, magnified to giant size, appeared in the middle of a triple rainbow! As we could no longer see the earth, all sensation of movement ceased. We might be going at storm speed and not know it. We could not even know the direction we were taking, save by descending below the clouds to regain our bearings!"

They knew they had been up in the air an hour when they heard the peal of church bells, the midday Angelus. Santos-Dumont, for whom every meal was a special occasion, declared that it was time for lunch. Machuron raised his eyebrows—he had not planned to descend so soon. But Santos-Dumont had no intention of returning either. With a mischievous look, he opened his valise and produced a sumptuous spread of hard-boiled eggs, roast beef, chicken, assorted cheeses, fruit, melting ice cream, and cake. To Machuron's delight, he also uncorked a bottle of champagne, which they thought was particularly effervescent due to the reduced air pressure at the high elevation. Santos-Dumont pulled out two crystal glasses. As he offered a toast to his host, he explained that he had never before eaten in such a splendid setting. The heat of the sun boiled the clouds, "making them throw up rainbow jets of frozen vapor like giant sheaves of fireworks. . . . Lovely white spangles of the most delicate ice formation scatter here and there by magic, while flakes of snow form moment by moment out of nothingness, beneath our very eyes, and in our very drinking glasses!" No dining experience was complete for Santos-Dumont without an after-dinner liqueur and fine Brazilian coffee, which he carried in a thermos.

While the two aeronauts sipped Chartreuse, the very snow that was entertaining them was quietly building up on top of the balloon. At least Machuron was sober enough to keep checking the instruments. At one point the barometer shot up five millimeters, signaling that the balloon, weighted down by the precipitation, must be falling rapidly even though they could not feel any movement. Suddenly they were plunged into half-darkness as the balloon passed through a cloud. They could still make out the basket, the instruments, and the parts of the rigging that were nearest to them, but the balloon itself was invisible. "So we had for a moment the strange and delightful sensation of hanging in the void without support," wrote Santos-Dumont, "of having lost our last ounce of weight in a limbo of nothingness." They furiously threw out ballast. After a few minutes they emerged from the dark fog to find themselves only one thousand feet above a village—the balloon had plunged nine thousand feet. The two men took their bearings with a compass and compared the landmarks they saw with those on a map. "Soon we could identify roads, railways, villages, and forests," Santos-Dumont said, "all hastening toward us from the horizon with the swiftness of the wind itself!" The wind was also gusting unpredictably, tossing the balloon from side to side and bouncing it up and down, making a soup of what remained of the roast beef and ice cream.

If this maiden voyage had taught Santos-Dumont the value of ballast in maintaining a balloon's equilibrium, it also taught him the value of the guide rope for a smooth landing and takeoff. The thick guide rope, extending three hundred feet and dangling from the basket, served as an automatic brake whenever the balloon returned to earth with disquieting speed for whatever reason. And the reasons could be many: a downward stroke of wind, the accidental loss of gas, the accumulation of snow on the balloon envelope, or a cloud passing in front of the sun. When the balloon descended below three hundred feet, more and more of the guide rope came to rest on the ground, thereby lightening the weight of the craft and arresting its fall. Under the opposite condition, when the balloon was ascending too rapidly, the lifting of the guide rope off the ground increased the weight of the balloon, thereby slowing its rise.

The guide rope, though ingeniously simple and effective, also had its "inconveniences," as Santos-Dumont charitably put it. "Its rubbing along the uneven surfaces of the ground— over fields and meadows, hills and valleys, roads and houses, hedges and telegraph wires—gives violent shocks to the balloon," he wrote later. "Or it may happen that the guide rope, rapidly unraveling the snarl in which it has twisted itself, catches hold of some asperity of the surface, or winds itself around the trunk or branches of a tree." He was writing from experience. As Machuron prepared to land, the guide rope coiled itself around a large oak, bringing the balloon to an abrupt halt, throwing the two aeronauts backward in the basket. For a quarter of an hour, the captured balloon, battered by the wind, kept them "shaking like a salad basket."

Machuron used the occasion to dissuade Santos-Dumont from constructing a powered balloon. "Observe the treachery and vindictiveness of the wind!" he shouted. "We are tied to the tree, yet see what force it tries to jerk us loose!" At that moment Santos-Dumont was thrown again into the bottom of the basket. "What screw propeller could hold a course against it?" Machuron continued. "What elongated balloon would not double up and take you flying to destruction?"

They eventually managed to free themselves from the oak by throwing out most of the remaining ballast. But the adventure was not over. "The lightened balloon made a tremendous leap upward," recalled Santos-Dumont, "and pierced the clouds like a cannonball. Indeed, it threatened to reach dangerous heights, considering the little ballast we had remaining in store for use in descending." The experienced Machuron had one last trick: He opened the balloon's valve to let gas escape and the balloon began to descend again toward an open field, the guide rope behaving itself this time as it came in contact with the ground. The field would normally have been an ideal landing spot, but a strong crosswind promised a harsh touchdown in so open an area. Fortune smiled on Santos-Dumont, though, and as the balloon fell, after nearly two hours aloft, it drifted toward the edge of the field.

"The Forest of Fontainebleau was hurrying toward us," Santos-Dumont recalled. "In a few moments we had turned the extremity of the wood, sacrificing our last handful of ballast. The trees now protected us from the violence of the wind; and we cast anchor, at the same time opening wide the emergency valve for the wholesale escape of gas." They landed smoothly, without any damage, climbed out of the basket, and watched the balloon expire. "Stretched out in the field, it was losing the remains of its gas in convulsive agitations," Santos-Dumont said, "like a great bird that dies beating its wings." And they could not have found a better deathbed, the well-manicured grounds of the Château de la Ferrière, owned by Alphonse de Rothschild, the seventy-year-old head of the Bank of France, the man responsible for the wealth of his famous family. Servants and laborers helped the two aeronauts fold up the collapsed balloon; stuff it, the rigging, and the lunch-table settings into the basket; and transport all 440 pounds to the nearest railway station two and a half miles away. On the sixty-mile train trip back to Paris, Santos-Dumont told Machuron that aeronautics was his calling. The balloon maker promised to build him his own pear-shaped balloon. Santos-Dumont's only disappointment was that he would have to put on hold his dream of a steerable airship. The two men and their balloon were back in Paris by 6:00 P.M. Santos-Dumont pronounced the day a success and started thinking about what he was going to have for dinner.

Machuron and Lachambre never had such an eager client nor a more contrary one. He was back in their workshop the next day placing an order for his first balloon, called *Brazil*. The misunderstandings began at once. Machuron assumed that he wanted an ordinary-size balloon that could hold between 17,000 and 70,000 cubic feet of gas. But Santos-Dumont had in mind a gasbag four times smaller than had ever been flown before, a balloon so compact, twenty feet in diameter, that when its 4,000 cubic feet of gas was released he could carry it around town in his handbag. Machuron refused to accept the order. He spent the afternoon trying to convince Santos-Dumont that *Brazil* would never fly.

"How often have things been proved to me impossible!"

Santos-Dumont wrote later. "Now I am used to it, I expect it. But in those days it troubled me. Still I persevered."

Machuron and Lachambre insisted that for stability a balloon had to have a certain minimum weight. The aeronaut needed the freedom to move around in the basket without fear that his actions would cause the balloon to rock or swing uncontrollably. With a small balloon, they said, that freedom would be impossible. Not so, Santos-Dumont argued. If the suspension tackle that connected the basket to the balloon were made proportionately longer, the center of gravity of even a lightweight system would not shift appreciably when the aeronaut moved about. He drew two diagrams to illustrate. The veteran balloon makers conceded that he had a point and made plans to construct *Brazil* from the usual materials.

Santos-Dumont had a problem with this too. The customary materials were too heavy, he said. He wanted to make the balloon out of light Japanese silk, and he brought Machuron a sample. "It will be too weak," Machuron said. "It will not be able to withstand the enormous pressure of the gas." Santos-Dumont wanted proof, so Machuron measured the strength of the silk with a dynamometer. The result surprised both of them. The silk was thirty times stronger than it needed to be. Although a square meter of silk weighed only a little more than an ounce, it could withstand a strain of more than 2,200 pounds.

By the time Santos-Dumont left the workshop, Machuron and Lachambre were shaking their heads. He had managed to persuade them to change every material that they ordinarily used. The silk envelope of the balloon would weigh less than four pounds. Three coats of varnish, to keep gas from seeping through, would bring the weight to thirty-one pounds. The netting that covered the balloon would be four pounds instead of hundreds of pounds, and the basket would weigh only thirteen pounds, five times lighter than usual.

Because Machuron and Lachambre had several orders to fulfill before they could start on *Brazil*, Santos-Dumont would have to wait a few months before he could see whether his lightweight balloon was flight-worthy. The two balloon builders were also booked for public ascents at fairs, festivals, and weddings throughout France and Belgium. Santos-Dumont preferred that Machuron and Lachambre remain in the workshop building *Brazil*, and so they agreed that he could ascend in their place after two training flights with Machuron. "As I got the pleasure and the experience, and paid all my expenses and damages, it was a mutually advantageous arrangement," said Santos-Dumont. All and all, he made more than two dozen flights before *Brazil* was completed.

On a stormy afternoon in March 1898, he filled in for Lachambre at a fair in Péronne, in the north of France. Thunder was rumbling in the distance, and some of the onlookers who knew that he was inexperienced urged him not to ascend at all or certainly not without a copilot. The expressions of concern made him more determined to go up in the balloon by himself.

"I would listen to nothing," he recalled. He went up late in the afternoon, as he had originally planned. "Soon I had cause to regret my rashness," he said. "I was alone, lost in the clouds, amid flashes of lightning and claps of thunder, in the rapidly approaching darkness of the night. On, on I went tearing in the blackness. I knew I must be going with great speed, yet felt no motion. I heard and felt the storm. . . . I felt myself in great danger, yet the danger was not tangible." He stayed up all night, waiting for the storm to break. The longer he waited, with no discernible damage to the balloon, the less fearful he was. "There was a fierce kind of joy," he said. "Up there in the black solitude, amid the lightning flashes and the thunderclaps, I was a part of the storm."

Once the bad weather passed, the joyous thrill of night ballooning turned to bliss. "In the black void," he said,

one seems to float without weight, without a surrounding world, a soul freed from the weight of matter! Yet, now and again there are the lights of earth to cheer one. You see a point of light far on ahead. Slowly it expands. Then where there was one blaze, there are countless bright spots. They run in lines, with here and there a brighter cluster. You know that it is a city. . . . And when the dawn comes, red and gold and purple in its glory, one is almost loath to seek the earth again, although the novelty of landing in who knows what part of Europe affords still another unique pleasure. . . . There is the true explorer's zest of coming on unknown peoples like a god from a machine. "What country is this?" Will the answer come in German, Russian, or Norwegian?

On this occasion the answer came in Flemish, because Santos-Dumont had landed far inside Belgium.

As soon as he returned to Paris, he urged his young male friends whose lust for adventure had been snuffed by the demands of family and business to take up ballooning. "At noon you lunch peacefully amid your family," he said. "At 2:00 P.M. you mount. Ten minutes later you are no longer a commonplace citizen—you are an explorer, an adventurer of the unknown as truly as those who freeze on Greenland's icy mountains or melt on India's coral strands." And the adventure did not always end with the landing. Other aeronauts, he told his friends, had been shot at when they descended in foreign countries. Some had been taken prisoner "to languish as spies while the telegraph clicked to the far-off capital, and then to end the evening over champagne at an officer's enthusiastic mess. Still others have had to strive with the dangerous ignorance, and superstition even, of some remote little peasant population. These are the chances of the winds!"

Santos-Dumont chose to make his first ascent in *Brazil* on July 4, 1898, at the Jardin d'Acclimatation, the zoological gardens in the Bois de Boulogne. The Bois was a huge wooded park, with two and a half times the acreage of New York City's Central Park. Earlier in the century it had been the stalking ground of thieves and ruffians. Napoleon III asked Baron Haussmann to redesign the Bois along the lines of London's Hyde Park. He turned some of the woods into open fields and added policemen, bungalows, pavilions, landscaping, and roads wide enough for horse-drawn carriages to make a U-turn. By Santos-Dumont's day, the Bois was the playground of the rich, with its neat polo grounds and the Longchamp horse-racing track.

The Jardin d'Acclimatation at the north end of the Bois opened in 1856. It was originally conceived as a scientific research center where animals of interest to French breeders would be acclimatized. Among the first inhabitants were yaks from Tibet, porcupines from Java, water pigs from South America, zebus from India, and zebras, kangaroos, cheetahs, llamas, ostriches, and armadillos. There were also Spanish mastiffs, Siberian greyhounds, and other dog breeds. Santos-Dumont's new friend Alphonse de Rothschild was a director of the Jardin d'Acclimatation, but the operation proved too expensive to run as a scientific venture and so by 1865 it was turned into a tourist destination with the introduction of crowd-pleasing zoo animals such as bears, elephants, hippo-potamuses, and dromedaries. Children could ride a train towed by a zebra or watch a car pulled by llamas that had a monkey as a coachman. But the zoo's directors were not content with showing animals. In the interest of drawing even more spectators, they decided to present living people too, "from the four corners of the world." American Indians, Eskimos, Nubians, Hindus, and Kurds were exhibited, complete with labels and maps of their range, as if they were exotic apes. On Sunday fashionably dressed women and their escorts strolled through the zoological gardens and gawked at the natives on display.

Santos-Dumont could have ascended from a more secluded spot, but he had confidence in *Brazil* and wanted to show it off to the many curiosity seekers in the Jardin d'Acclimatation. Because Lachambre had built a hydrogen plant there, it was also a convenient place. The little balloon with disproportionately long rigging proved equal to the challenge. Santos-Dumont showed unusual restraint in leaving behind his substantial lunch basket so that *Brazil* could hold the maximum amount of ballast, sixty-six pounds of sand. Although *Brazil* contained only one-seventh the hydrogen gas of a typical balloon, it easily carried him and the ballast aloft. As Machuron and Lachambre anxiously watched from the ground, he demonstrated *Brazil*'s stability by making a big show of moving around in the basket. Relieved, the two men helped themselves to a bottle of champagne that he had left behind. After the smooth descent, Santos-Dumont pulled the rip cord, waited a short while for the balloon to deflate, and packed the whole thing into his valise.

The flawless flight gave him confidence. If the veteran aeronauts had misjudged the stability of *Brazil* and underestimated the strength of Japanese silk, could they not also be wrong about the difficulties of building a steerable balloon? How could they be so sure that a propeller-driven airship would collapse in a strong wind? What if he changed the shape of the balloon from a near sphere to an elongated cylinder? Instead of being bandied about by the wind, would it not "cut the air"?

What eluded him at first was the power source. The petroleum engine was an unlikely candidate because it was unreliable as well as deafening and foul-smelling—characteristics that would detract from the tranquillity of ballooning. Petroleum engines in automobiles seemed to have minds of their own, slowing down and speeding up and conking out at will, which was bad enough if you had a road under you but unacceptable in the air.

Santos-Dumont had acquired half a dozen automobiles since his Peugeot roadster. Although he was not satisfied with their performance, he enjoyed taking the sputtering vehicles for a spin. His notion of an autumn holiday was driving a six-horsepower Panhard six hundred miles from Paris to Nice; he made it in fifty-four hours, stopping often to make minor repairs and tweak the engine but not to sleep. He never made a long road trip again, however, because he could not stand to be away from his balloons.

Eventually he had even stopped using his cars for everyday driving. "I was once enamored of petroleum automobiles because of their freedom," he told a journalist some years later. "You can buy the essence everywhere: and so, at a moment's notice, one is at liberty to start off for Rome or St. Petersburg. But when I discovered that I did not want to go to Rome or St. Petersburg, but only to take short trips about Paris, I went in for the electric buggy," of a kind seldom seen in France.

In 1898, he imported a light electric vehicle from Chicago and "never had cause to regret the purchase." Every day he went for a morning spin through the gardens of the Bois and on afternoon errands to the balloon makers' workshop in Vaugirard and the Automobile Club in the place de la Concorde. The electric motor, aside from its reliability, had other advantages over the petroleum engine: It was quiet and odorless. But it was not suitable for air travel because, with its required batteries, it was much too heavy, and modifying it seemed unlikely. He knew that the French government in supporting Renard and Krebs's efforts in the 1880s "had spent millions of francs on air-ships with electric motors whose plan had finally been abandoned chiefly because of the motor's weight."

One day, as he rode around Paris on a motorized De Dion tricycle, it occurred to him that he may have prematurely dismissed the petroleum motor. The single-cylinder tricycle engine, he realized, "happened to be very much perfected at the moment," compared with the troublesome higher-powered petroleum engines in four-wheeled automobiles. Pound for pound the 1.75-horsepower motor in his De Dion was relatively powerful, although not strong enough to guide an airship. To increase the power, he planned to combine two of them. Usually he was cocky about his inventions, but this time he was not confident enough to experiment in public.

"I looked for the workshop of some little mechanic . . . in the central quarter of Paris," recalled Santos-Dumont. "There I could have my plans executed under my own eyes and apply my own hands to the work. I found such a workshop in the Rue du Colisée. There I worked out a tandem of two cylinders of a petroleum motor, that is, their prolongation, one after the other, to work the same connecting rod, while fed by a single carburetor. To bring everything down to the minimum of weight, I cut out from each part what was not strictly necessary to solidity. In this way I realized something which was remarkable at the time—a 3½ horse-power motor weighing only sixty-six pounds."

He was pleased with his handiwork and set out to test the reconstructed engine in his tricycle. The Paris–Amsterdam automobile race was approaching, and he could not think of a better way to put the engine through its paces than to enter the competition. He was disappointed to learn that his souped-up vehicle did not meet the eligibility requirements but made the best of the situation by driving the tricycle alongside the race until he convinced himself that he could keep pace with the leaders. "I might have had one of the first places at the finish (the average speed was only 40 kilometers, or 25 miles per hour)," wrote Santos-Dumont, "had I not begun to fear that the jarring of my motor in so long and strenuous an effort might at last derange it and delay the more important work on my air-ship. I, therefore, fell out of the race while still at the head of the procession."

The shaking of the motor reminded him of how the machines on the coffee plantation had fallen apart from their own vibrations. To ensure that his balloon engine would not have a similar fate, he drove his tricycle to the Bois in the middle of the night when the park was abandoned. He had hired two burly workmen to meet him there with heavy-duty ropes and paid them generously so that they would tell no one about the nocturnal experiments. He selected an ample tree with a thick branch just above his head. The workmen tossed the ropes over the branch and tied them securely to both ends of the tricycle. He mounted the vehicle and gave the order to hoist him five feet into the air. With the engine going full throttle, he sat there feeling the vibrations; they were noticeable but were much less than they were on the ground, where the engine had something to vibrate against. He pronounced the test a success, swore the workmen to secrecy once more, and sneaked out of the park before he could be arrested for violating the curfew.

When dawn broke, he told friends about his plan. "From the beginning everybody was against the idea," he recalled. "I was told that an explosive gas engine would ignite the hydrogen in the balloon above it, and that the resulting explosion would end the experiment with my life." He reminded his doubters that half a century earlier Henri Giffard had gone up in a hydrogen balloon powered by a fiery steam engine and although the flight was only a qualified success (because the engine was not powerful enough to work against the wind), Giffard made it down unscathed. The tricycle engine, Santos-Dumont insisted, would spit far fewer sparks and less smoke.

He wrote down his plans for a cigar-shaped airship and returned to the balloon makers in Vaugirard. When he tried to place an order for the balloon, Lachambre at first refused to take it, Santos-Dumont recalled, "saying that such a thing had never been made, and that he would not be responsible for my rashness." Santos-Dumont reminded him that he had voiced similar doubts before building *Brazil*. He also promised to indemnify Lachambre against any explosion or damages and agreed to work on the engine himself far away from the workshop. Worn down by Santos-Dumont's persistence, Lachambre "went to work without enthusiasm."

Santos-Dumont's guiding principle in designing the dirigible, which he called *Santos-Dumont No. 1* in anticipation of building a series of airships, was to make it the smallest elongated balloon that could carry aloft the reconfigured engine, a propeller, a rudder, the balloon basket, the rigging, a minimum of ballast, the guide rope, and of course his own weight, which fluctuated between a hundred and a hundred ten pounds depending on how well he had been eating. The sketch he gave Lachambre called for "a cylinder terminating at each end by a cone." It would be 82.5 feet long by 11.5 feet in diameter, with a gas capacity of 6,454 cubic feet, which would give it a lifting power of 450 pounds. Knowing the weight of everything the balloon would have to take aloft, Santos-Dumont computed that only sixty-six pounds remained for the balloon material, the varnish, and the chemise (the outer cover, or netting, by which the balloon was attached to the basket). The use of Japanese silk, which had proved so staunch in *Brazil*, would not by itself keep the weight within sixty-six pounds. He needed another innovation. First he considered alternatives to the varnish, but he could not find a lighter liquid that would sufficiently seal the silk. Then he focused on the chemise, and ended up doing away with it entirely. The rigging lines from the basket would now be attached directly to the balloon envelope, the lines joined to long wooden rods housed in horizontal hems stitched into the balloon fabric. Santos-Dumont was proud of this simple idea, and instructed a reluctant Lachambre to sew the balloon accordingly. The veteran balloon maker was worried that the stitches might rip, releasing the basket for a fatal free-fall. As with the engine, Santos-Dumont absolved him of all responsibility.

While Lachambre went to work, Santos-Dumont busied himself in the rue du Colisée workshop getting the motor ready. He switched the motor from his tricycle to the rear of the balloon basket and attached an aluminum propeller directly to the motor shaft. By suspending the basket, the motor, and the 6.6-foot propeller from the rafters of the workshop, he got an idea of how the machinery would perform in the air. With the motor full throttle, the basket shot violently forward. Pulling the basket back with a horizontal rope attached to a dynamometer, Santos-Dumont measured the traction power of the propeller to be as high as twenty-five pounds, "promising good speed for a cylindrical balloon of my dimensions, whose length was equal to seven times its diameter." He repeated the trials daily to be sure of the results. If all went well, he concluded, the airship would cruise along at eighteen miles an hour.

Having introduced the propeller (and a rudder made from silk stretched over a triangular steel frame) to wrest control of the balloon's horizontal motion from the vagaries of the wind, he turned his attention to the question of vertical equilibrium, which was uneasily maintained in spherical balloons by the jettisoning of ballast or the venting of gas. "Suppose you are in equilibrium at five hundred meters height," he wrote.

All at once a little cloud, almost imperceptible, masks the sun for a few seconds. The temperature of the gas in your balloon cools down a little; and if, at the very moment, you do not throw out enough ballast to correspond to the ascensional force lost by the condensation of the gas, you will begin descending. Imagine that you have thrown out the ballast—just enough, for if you throw too much, you will become too light and go too high. The little cloud ceases to mask the sun. Your gas heats up again to its first temperature and regains its old lifting-power. But, having less to lift by the amount of ballast thrown out, it now shoots higher into the air, and the gas in the balloon dilates still more, and either escapes through the safety-valve or has to be deliberately sacrificed, and the trouble recommences. These *montagnes-russes,* or shoot-the-chutes, vagaries of spherical ballooning must be avoided to the utmost with my air-ship.

It occurred to Santos-Dumont that his new propeller might give him control of his elevation if he could figure out how to tilt the airship, raising or lowering its nose, so that the motor would drive the balloon's ascent or descent. Once again his solution was simple: a system of movable weights by which the center of gravity of the airship could easily be shifted. The weights were merely two bags of ballast, one fore and one aft, suspended from the balloon envelope by long, heavy cords. Extending from each weight to the basket was a lighter cord, by which the weight could be pulled into the basket, shifting the center of gravity of the whole system. If the front weight was drawn in, the nose of the airship would point up, and if the aft weight was pulled in, the nose would point down. Other than the two-hundred-foot guide rope, which would be useful during takeoff and landing, *No. 1* would not require additional ballast. Santos-Dumont hoped that he had minimized the ballast enough to have weight to spare for an ample lunch basket. He was ready to fly *No. 1* as soon as Lachambre applied the varnish.

On September 18, 1898, three and a half months after his first ascent in *Brazil,* Santos-Dumont put *No. 1* to the test. By then, he had floated over Paris more than a hundred times in spherical balloons, and his reputation as a courageous and ingenious balloonist was known throughout the city. No other Parisian aeronaut was flying for his own pleasure; the others were paid professionals, and they ascended mostly in rural areas. He had already earned the nickname Petite Santos. It was meant affectionately but it bothered him. He had gone to great lengths—dark, vertically striped suits, lifts in his shoes,

a panama hat—to boost his short stature. He had even designed custom-made high collars for his dress shirts to make
his neck appear longer. He drew the knot in the tie excessively
tight so as not to accentuate his small size, preserving the
tightness by piercing the knot with a pearl or jeweled pin. His
suit jacket and turned-up trousers were always crisply pressed.
He was the most impeccably dressed aeronaut the world would
ever know.

People turned out at his ascensions as much to see him as
to watch him fly. The accessories to his wardrobe were decidedly feminine and piqued the interest of spectators and journalists alike, who could not reconcile them with his manly
risk-taking in novel airships. One foreign correspondent described him this way:

> Santos, as he prefers to be called, is a little, thin, swarthy
> chap of 5 feet and maybe 4 or 5 inches. His face would
> be effeminate were it not for the thick, though closely
> cropped, mustache, which shades his upper lip, and lends
> strength to his whole face. His chin shows, however,
> whence he gets the dogged sticktoitiveness and the won-
> derful grit which has enabled him to keep on working until
> at last he has reached his present eminence. The lower
> jawbone is long and angular, and when he closes it the
> protrusion of the muscles denoting determination is very
> pronounced. The roof of his mouth is inclined to protrude
> also, and his lips are a trifle thicker than the average. He
> is not a handsome man. His teeth are, however, beautifully
> white and regular, and his smile is charming. It spreads all
> over his face, beginning with his eyes, and as it steals over
> his features it softens and lightens them delightfully. . . . It
> is his voice, too, which is low and strangely gentle, which
> somehow conveys the idea of effeminacy which one cannot

help but feel no matter how often one is reminded of his daring feats of courage. This effect is added to by a gold bracelet which Santos wears on his wrist, although his sleeve hides it, except occasionally, when some gesture of the arm shows it for a moment. This is rare, however, for Santos thinks much more than he talks, and talks much more than he gestures.

Fellow aeronauts and members of the Automobile Club turned out early at the Jardin d'Acclimatation on September 18 to watch him prepare *No. 1*. The zoological gardens were home now to one of Lachambre's large tethered balloons. Lachambre sold hydrogen to him at the favorable rate of one franc per cubic meter, the gas for *No. 1* costing $1,270. As Santos-Dumont inflated the airship, the assembled aeronauts spoke nervously among themselves. Finally one of them shared their concern about the potentially lethal combination of a fire-spitting motor and a highly flammable gas: "If you want to commit suicide, why not sit on a cask of gunpowder with a lighted cigar in your mouth?"

Santos-Dumont laughed and assured the onlookers that he most decidedly wanted to live, if only to witness the future of flying machines. He pointed to the exhaust pipe on the engine. He proudly showed them how he had bent the pipe with his own hands so that any sparks were directed away from the balloon. Besides, he said, he was so familiar with the tricycle engine that he could tell by subtle changes in its sound if it was starting to burn uncontrollably, in which case he would just shut it off.

The issue of the motor was abandoned, however, when the bystanders saw him doing something that looked far more perilous: He was preparing to start his ascension at the downwind end of the open turf, next to the woods. Although the airship was facing upwind, many assumed that the motor would not be a match for the wind and that he would be swept backward a few feet into the nearby trees. Santos-Dumont was convinced that his motor was more powerful than the wind. He planned to adjust it until the force of the propeller exactly canceled out the wind, so that the balloon would rise straight up. The other aeronauts pleaded with him not to make such a risky takeoff on the first flight. Why not adopt the time-honored approach in spherical ballooning of starting the ascent at the upwind end of the open space? That way the balloon, pushed by the wind as it climbed, would have the entire expanse to cross before reaching the woods. Santos-Dumont gave in to the crowd and moved *No. 1* to the other end of the field. It was the wrong approach.

He aimed the airship downwind across the open field and, with the engine idling, climbed into the basket. Then he shouted, "Let go all!" Machurin and Albert Chapin, Santos-Dumont's chief mechanic, released the mooring ropes and the Brazilian throttled up the engine. *No. 1* raced forward, propelled by the wind and motor working together, and in a matter of seconds traversed the field and smashed into the trees on the other side. "I had not time to rise above them," Santos-Dumont recalled, "so powerful was the impulse given by my motor." The airship fell to the ground—fortunately the descent was cushioned by the scraping of the basket along the branches—and he emerged with no injuries except to his pride. He berated his fellow aeronauts for talking him out of his plan. Never again, he said, would he have the "weakness to yield." But the episode had its dividends. "This accident," he said, "at least served to show the effectiveness of the petroleum motor in the air to those who doubted it before."

In two days he repaired the airship and returned to the Jardin d'Acclimatation for a second attempt. The crowd was larger this time, drawing strangers who were torn between fear and excitement that they might witness another crash. There was a stiff breeze, and this time Santos-Dumont stuck to his instinct of positioning the airship at the downwind end of the lawn and aiming it into the wind. *No. 1* rose slowly and was never in danger of crashing into the trees. He reeled the front ballast weight into the basket, and, with the center of gravity shifted toward the back, the balloon's giant nose swung up-ward. The crowd cheered. He tipped his hat and began to demonstrate that he could indeed steer the balloon. He grasped the rudder and guided *No. 1* in a tight loop around Lachambre's captive balloon. The applause was even louder, and Lachambre saluted his protégé.

Santos-Dumont's first surprise was that he could actually sense the airship moving, unlike the experience in a spherical balloon. He was astonished to feel the wind in his face and his coat fluttering as *No. 1* plowed ahead. He likened it to standing on the deck of a fast-moving steamship. He had won-dered whether the sensation of mounting and descending obliquely with his shifting weights would be unpleasant. But it turned out not to bother him at all even though *No. 1* pitched considerably. He attributed his composure to sea legs earned on voyages between France and Brazil. "Once, on the way to Brazil," he recalled,

> the storm was so violent that the grand piano went loose and broke a lady's leg; yet I was not seasick. . . . I know that what one feels most distressingly at sea is not so much the movement as that momentary hesitation just before the boat pitches, followed by the malicious dipping or mount-

ing, which never comes quite the same, and the shock at top and bottom. All this is powerfully aided by the smells of the paint, varnish, and tar, mingled with the odors of the kitchen, the heat of the boilers, and the stench of the smoke and the hold. In the air-ship there is no smell. All is pure and clean. And the pitching itself has none of the shocks and hesitations of the boat at sea. The movement is suave and flowing, which is doubtless owing to the lesser resistance of the airwaves. The pitches are less frequent and rapid than those at sea; the dip is not brusquely arrested, so that the mind can anticipate the curve to its end; and there is no shock to give that queer "empty" sensation to the solar plexus.

And the navigator of the air, he observed, has one great advantage over the sea captain—he can easily move laterally to trade an undesirable current for an advantageous one.

At first the flight of *No. 1* could not have gone better. "For a while we could hear the motor spitting and the propeller churning the air," reported an eyewitness. "Then, when he had reached equilibrium, we could still observe Santos manipulating the machinery and the ropes. Around and around he maneuvered in great circles and figure 8's, showing that he had perfect control of his direction."

Santos-Dumont was encouraged by the ease with which he controlled *No. 1*. "Being inexperienced," he said, and over-confident, "I made the great mistake of mounting high in the air—some 1300 feet—an altitude that is considered nothing for a spherical balloon, but which is absurd and uselessly dangerous for an air-ship under trial." At that height he commanded a view of the entire city and was enthralled by the beautiful grounds of Longchamp. He headed toward the racetrack.

"As the air-ship grew smaller in the distance, those who had opera-glasses began crying that it was 'doubling up,' " the eyewitness continued. "We saw it coming down rapidly, growing larger and larger. Women screamed. Men called hoarsely to one another. Those who had bicycles or automobiles hastened to the spot where he must be dashed to the ground. Yet within an hour M. Santos-Dumont was among his friends again, unhurt, laughing nervously, and explaining all about the unlucky air-pump."

He told his friends that he had encountered no problems when he ascended. As the atmospheric pressure decreased, the hydrogen simply expanded, keeping the balloon taut. And when the expansion became too great, a valve automatically released some of the gas. The valve was another one of Santos-Dumont's innovations. Spherical balloons generally had a free vent, a small open hole, in the bottom through which gas could escape as it expanded. The free vent meant that there was never any danger of the balloon bursting, "but the price paid for this immunity," he noted, "is a great loss of gas and, consequently, a fatal shortening of the spherical balloon's stay in the air." And it was not just an issue of prolonging the flight that was on his mind when he substituted a valve for the open hole. He was also concerned about maintaining the airship's cylindrical shape. When a spherical balloon lost a bit too much gas, it had a limp shape but was still flightworthy. If his cylindrical balloon leaked gas, it started to fold and was difficult, if not impossible, to fly. The introduction of the valve eliminated the accidental leaking of gas, but its proper functioning was critical to his safe return. He repeatedly checked the valve just before the trip, because, although his friends saw fire as the chief danger, his principal concern was the valve failing and the balloon exploding.

But on the actual flight the problems occurred on the descent. The increase in atmospheric pressure compressed the balloon, as he had expected. He had equipped *No. 1* with an air pump that was supposed to direct air into the balloon to compensate for any contraction. That was the idea anyway, but in practice the pump proved to be too weak.

As Santos-Dumont descended, *No. 1* began to lose its shape, folding in the middle like a portfolio. The cords were subjected to unequal tension, and the balloon envelope was in danger of being torn apart. "At that moment I thought that all was over," recalled Santos-Dumont, "the more so as the descent which had already become rapid could no longer be checked by any of the usual means on board, where nothing worked." The cords suspending the ballast bags became tangled, so he could no longer control where the nose of the airship pointed. He thought of throwing out ballast. That would certainly cause the airship to rise, and the decreased atmospheric pressure would enable the expanding hydrogen gas to restore the balloon to its taut, cylindrical shape. But when he eventually returned to earth, the problem would undoubtedly repeat itself, only worse; the balloon would be flaccid because of the gas lost in the interim. Santos-Dumont could think of nothing to do as *No. 1* plummeted. He feared that the cords connecting the basket to the balloon would snap one by one. He looked down, and the sight of the housetops, "with their chimney pots for spikes," made him queasy.

"For the moment," he wrote, "I was sure that I was in the presence of death. . . . 'What is coming next?' I thought. 'What am I going to see and know in a few minutes? Whom shall I see after I am dead?' The thought that I would be meeting my father in a few minutes thrilled me. Indeed, I think that in such moments there is no room for either regret or terror. The mind is too full of looking forward. One is frightened only so long as one still has a chance."

But then he realized that he did have a chance. A charitable wind was sweeping him away from the rocky streets and jagged roofs toward the soft grassy *pelouse* of Longchamp, where a few boys were flying kites. He shouted to them to grab the two-hundred-foot guide rope and run with it as fast as they could against the wind. "They were bright young fellows," he recalled, "and they grasped the idea and the guide-rope at the same lucky instant. The effect of this help *in extremis* was immediate, and such as I had expected. By this maneuver we lessened the velocity of the fall, and so avoided what would otherwise have been a bad shaking-up, to say the least. I was saved for the first time!" The boys helped him pack everything into the airship's basket. He secured a cab and returned to the center of Paris.

He immediately put the troublesome aspects of the flight behind him, like a mother forgetting the pains of labor once she has seen her newborn's face. "The sentiment of success filled me," he recalled. "I had navigated the air. . . . I had mounted without sacrificing ballast. I had descended without sacrificing gas. My shifting weights had proved successful, and it would have been impossible not to recognize the capital triumph of these oblique flights through the air. No one had ever made them before."

That night he celebrated at Maxim's, the famous restaurant at No. 3, rue Royale that is still in business today. He was one of Maxime Gaillard's first customers, when the dark-wooded bistro opened in the early 1890s. The restaurant in-itially catered to carriage drivers who passed the time while their bosses dined elsewhere, but soon they too discovered its fine, hearty cuisine—French onion soup, oysters on the half shell, poached lobster, sole in brandy sauce, roast chicken, scallops of veal, grilled pigs' feet and tails—and displaced the coachmen. As a night spot for the well-to-do, Maxim's was ideally located in the center of the city, on the same block as the Automobile Club, the aristocratic Hotel Crillon, and the elite Jockey Club. Maxim's attracted what working-class Parisians derisively called *des fils à papa*, rich young men who spent their fathers' money on women and wine. When it came to the wine, Santos-Dumont fit right in. Maxim's did not serve lunch in those days. The restaurant opened at 5:00 P.M. for the evening aperitif, dinner was served from 8:00 until 10:00, and supper from midnight until dawn.

Santos-Dumont always came for supper and sat at the same table in the corner of the candlelit main room. With his back to the wall, he could watch everything that transpired, and the goings-on in the wee hours of the night were legendary. A beautiful blonde who became a silent movie star used to shed all her clothes, climb onto one of the tables, and sing torch songs. A Russian named Aristoff arrived every morning precisely at four and consumed the identical meal: grilled kipper, scrambled eggs, minute steak, and a bottle of champagne. For his bachelor party, a French count ordered the waiters to dress up as undertakers and arrange the tables to looks like funeral biers. Maxim's was the spark for many romantic assignations. In the 1890s strangers rarely approached each other directly but flirted with their eyes across the dining room. Many couples got together because of the intervention of the notorious "Madame Pi-Pi," who sat outside the bathrooms and cleaned the toilets after each use. A woman who was interested in a man would excuse herself to the toilet and slip Madame Pi-Pi her address or phone number along with a tip. When she returned, the man could go to the bathroom and pay Madame Pi-Pi for the information.

Santos-Dumont dined alone or with close friends like Louis Cartier and George Goursat, better known by his nom de plume Sem, who carved the Brazilian's likeness on the restaurant's wall. At Maxim's Santos-Dumont met James Gordon Bennett, the American millionaire publisher, who had the most prominent table in the front of the restaurant. Bennett owned the *New York Herald* and the *Paris Herald,* the only English-language daily in the city. He had an odd sense of humor, which infused his papers. For instance, he ordered the *New York Herald* to print the same letter to the editor day after day for seventeen years—an 1899 note from "Old Philadelphia Lady" who wanted to know how to convert centigrade temperatures into Fahrenheit—because he enjoyed hearing from readers who pointed out the repetition. Bennett was a fan of fast cars, slick yachts, and hot-air balloons. He assigned a reporter to cover every trial of Santos-Dumont's airships. The *Herald,* with its hundreds of cliff-hanging stories about his perilous flights, made Santos-Dumont a celebrity in the United States.

On the days when Santos-Dumont planned to fly, the kitchen at Maxim's packed him lunch. H. J. Greenwall, the author of *I'm Going to Maxim's,* described the Brazilian's routine: "Out to the hangar to get *Santos-Dumont I* tuned up for a flight; lunch put in the wicker undercarriage in which the pilot flew. Up in the air went *Santos-Dumont I;* usually some minor accident or incident occurred. Back to the hangar. Back to his apartment," at the fancy address of No. 9, rue Washington, on the corner of the Champs-Elysées, near the Arc de Triomphe. "Back to Maxim's . . . all night; leave in the dawn with a lunch of say a wing of cold chicken, a salad, and some peaches. A short sleep. Then to the hangar and up goes Santos-Dumont again."

[CHAPTER 4]

DYING FOR SCIENCE

PARIS, 1899

•

AT THE CLOSE of the nineteenth century, Santos-Dumont was the only person flying powered airships. (Count Ferdinand von Zeppelin in Germany was building a mammoth, 420-foot-long semirigid airship—a fabric-covered, aluminum-strutted hull that housed fifteen separate gasbags—but it had not yet gone aloft.) Santos-Dumont's fellow aeronauts were still ascending in spherical balloons, and not always successfully. In 1898, the London *Evening News* challenged balloonists to make it across the Channel from London to Paris. A man named A. Williams, after months of waiting for a favorable wind, planned to take off on November 22. When he was nearly ready, "a slight accident took place," the paper said, "which delayed matters, and the start was postponed for an hour." It seems that while the balloon was being inflated, it was somehow driven against iron railings and ripped. Once the tear was repaired and the inflation completed, Williams discovered that the balloon was not capable of lifting two companions, as he intended, so only one, a Mr. Darby, accompanied him. After an hour they descended into a tree and then briefly rose again.

Finally, after having traversed a distance of not one quarter of that from London to Paris (and, by the way, not in the right direction), it was found that the balloon had not sufficient power to proceed, and a descent was attempted near Lancing. Then it was found that this balloon, supposed to be replete with everything that practical aeronauts could suggest, had not a single anchor on board. The aeronaut, not wishing to be carried over the sea, then adopted the extraordinary course of swarming down the guide rope, leaving his unfortunate companion to follow. Relieved of Mr. Williams' weight, the balloon started to rise again, and the passenger found himself in the awkward predicament of either having to jump some 50 ft. down, or be wafted out to sea. He chose the former, and though badly injured, was lucky enough to escape with his life. The balloon disappeared over the Channel, but was found some days afterward in France.

Mr. Darby was lucky. In 1899, the publication *Revue Scientifique* counted nearly two hundred people who had lost their lives in balloons. Usually the deaths were nasty and quick. Each monthly issue of *Aeronautical Journal,* a British periodical that tracked developments in flight around the world, published an accident report. In October 1899, the journal described two fatal falls:

An Italian military captive balloon broke loose in July, carrying with it not only an officer and a corporal who were in the car, but also a soldier who had held on to the rope in hopes of keeping the balloon down. Those in the car tried to draw up the unfortunate man, but after a time he let go, and was dashed to pieces on the banks of the Tiber. . . .

At Beuzeville, in France, an aeronaut, named Bernard, made an ascent, but finding his balloon had too little lift, he dispensed with the car, and sat upon the hoop. It is supposed that the gas from the balloon issuing through the neck must have asphyxiated him, for he was seen to let go his hold and fall to the ground from a great height, being killed on the spot.

Early in his planning to power an airship with an internal combustion engine, Santos-Dumont learned that Karl Wolfert, a Protestant minister, had had the same idea. Wolfert sought out the technical advice of automotive pioneer Gottlieb Daimler. On June 12, 1897, before an audience of the kaiser's military advisers, Wolfert was set to ascend with Michael, his mechanic, and an officer of the Prussian army. Just before liftoff, the officer was overcome by a bout of claustrophobia in the balloon basket and bowed out of the trip. In Wolfert's eagerness to take off and impress his distinguished observers, he neglected to add ballast to compensate for the missing officer's weight. He and his mechanic ascended to exuberant cheers and waves from the crowd, and the poorly ballasted airship climbed rapidly to three thousand feet. Without warning, the gasbag exploded, and the airship was engulfed in flames. A horrible scream was heard, and then complete silence. The stunned audience scrambled out of their seats to avoid the falling, charred wreckage. Two bodies, burnt beyond recognition, smashed the seats. They had died in just the kind of accident that Santos-Dumont's friends had feared.

On May 12, 1902, Santos-Dumont witnessed a similar accident in Paris that claimed the life of a fellow Brazilian, Augusto Severo. Inspired by Santos-Dumont's own efforts, Severo had built an airship called *Pax*. On Severo's first free ascent, accompanied by Saché, his machinist, sparks from the engine ignited the balloon and the hydrogen exploded. The frame of the airship plunged fifteen hundred feet and struck the one-story house at No. 89, avenue du Maine, collapsing its roof into the bedroom of a man named Clichy. The bed was on the opposite side of the room from the falling debris, and Clichy and his wife were awakened to the sight of a smashed airship and two disfigured bodies crashing through the ceiling. The *Herald* reported, "The machinist lay near the motor, stretched out upon the willow framework, which served as the flooring. His face had been terribly burned and his hands stripped of skin. His back was broken by the shock. M. Severo, who seems to have been standing at the moment of the shock, had nearly all his bones broken. He was scarcely recognizable. The shin bones protruded through the skin, and the lower jaw was torn from the socket." Santos-Dumont was grief-stricken, but the grisly accident only reinforced his resolve to build a safe and reliable airship.

AERONAUTS WERE not the only martyrs for science at the end of the nineteenth century and the beginning of the twentieth. The pace of industrial and scientific progress was so exhilarating that men and women were willing to sacrifice their own well-being to ensure that the progress continued. Scientists had always known that there was a general risk in exploring the uncharted—it was an occupational hazard. But in fin de siècle Europe and America, the stakes were raised. As the prestigious new American journal *Science* announced in 1883: "Higher than all, [science] must be devoted to the truth. It must cheerfully undertake the severest labor to secure it, and must deem no sacrifice too great in order to preserve it." Science had become the new secular religion, and its practitioners, like the aeronauts, were expected to proceed with an important experiment even if it might kill them.

Physicians at that time had few reservations about experimenting on themselves. In *Who Goes First?*, Lawrence Altman told the story of the French doctors who developed the rabies vaccine (and whose reputation drew Santos-Dumont's ailing father to Paris). Transmitted by the bite or lick of an infected animal, rabies was a relatively rare disease but notorious because of its horrifying symptoms—the slow but fatal destruction of the brain and central nervous system that leaves the victim gasping for air and shaking spasmodically—and its painful treatment—"cauterization of the bite with a red-hot iron." In 1880, Louis Pasteur, who was already revered for his "pasteurization" of milk and beer, turned his attention to the disease. Within a year he had found a method of transmitting the virus, by injecting brain material extracted from a rabid dog into a healthy canine. Soon he developed a technique for treating the brain extract so that he could adjust the virulence of the inoculation. Rabies acquired through a bite had a long incubation period. By giving a bitten dog a series of progressively stronger inoculations, the animal would develop immunity to the disease before the incubation period was over. In 1884, Pasteur reported that twenty-three immunized dogs had warded off the disease, but he was still afraid of using the live vaccine on human beings. He rebuffed the emperor of Brazil, who had pleaded with him to apply the vaccine in a country where the incidence of rabies was much greater than in Europe.

"Experimentation permitted on animals," Pasteur said, "is criminal when it comes to man." Rabies was a fatal disease, so the failure of the vaccine promised almost certain death. In 1885, he told three of his colleagues that he wanted to test the vaccine on himself. He took off his shirt and begged them to inject him with the live virus. They refused. They did not want to be accomplices to the possible suicide of one of France's most beloved scientists. Instead, the three men assumed the risk themselves. When weeks passed and none of them contracted the disease, Pasteur had the confidence to inoculate victims of rabid-dog bites. By 1886, he had treated 350 people and all but one was saved from the disease.

Medical self-experimentation was not confined to France. In 1892, seventy-four-year-old Max von Pettenkofer, the German public-health advocate who had purified Munich's drinking water, purposely swallowed a solution of cholera bacteria. He believed that the bacteria could not by itself cause the often-fatal disease, that other cofactors he identified needed to be present as well. Because he personally did not possess the cofactors, his dramatic experiment was intended to prove that cholera bacteria was not the sole causative agent. He had diarrhea for a week but never became seriously ill, confirming for himself the validity of his theory (although science would ultimately prove him wrong and attribute his mild symptoms to immunity from an earlier, accidental exposure to cholera). Pettenkofer had prepared himself for the worst. "Even if I had deceived myself," he wrote, "and the experiment endangered my life, I would have looked Death quietly in the eye for mine would have been no foolish or cowardly suicide; I would have died in the service of science like a soldier on the field of honor. Health and life are, as I have so often said, very great earthly goods but not the highest for man. Man, if he will rise above the animals, must sacrifice both life and health for the higher ideals."

On November 8, 1895, the German physicist Wilhelm Conrad Roentgen discovered X rays in his laboratory in Würzburg. The discovery was serendipitous: Roentgen had been experimenting with a cathode-ray tube in his darkened laboratory when he noticed that metals and other materials far from the tube were emitting an eerie green fluorescence. He suspected that radiation from the tube was causing the materials to glow but it could not be the familiar, short-distance cathode rays because they would not have reached the materials. When he inadvertently passed his hand between the tube and a glowing screen, he saw the outline of his bones. He hurriedly "photographed" his wife's hand and announced his discovery to the world. The "penetrating" radiation captured the public's imagination. X rays were featured in advertisements, popular songs, cartoons, novels, and breathless newspaper reports.

"The x-ray mania began early and grew quickly," social historian Nancy Knight noted:

"Hidden Solids Revealed!" trumpeted the *New York Times* in January 1896. The press was enchanted with the possibilities of the new rays. With the information that they rendered "Wood and Flesh More Easily Penetrated . . . Than Plain Glass," many observers immediately speculated on various applications and uses. Even the most mundane experiments with the new technique were labeled miraculous. "Startling results" announced by professors at Yale turned out to be x-ray photographs of uncracked walnuts showing "a splendid view of the kernels." Some popular magazines and journals showed x-ray photographs of feet in boots, coins in wooden boxes, and shapely women in tight lacing. One popular cartoon hinted at the possible leveling effects of the rays by revealing that beneath the superficial layer the well-to-do of the Gilded Age were the same as the common people.

Well-fed or hungry, fat or thin, everyone's skeleton looked roughly the same. Another cartoon, called "The March of Science," showed an eavesdropper behind a door. The caption said, "Interesting result attained, with the aid of Röntgen rays, by a first-floor lodger when photographing his sitting-room door."

Even as the X-ray craze abated, physicians continued to be smitten with the invisible new light. Within two months of Roentgen's discovery, the medical community knew that X rays were a powerful tool for revealing the interior of the human body. Physicians welcomed X rays because the Industrial Revolution had largely passed them by. The nineteenth century had seen great advances in the prevention of disease (through vaccines, antiseptic practices, and public-health initiatives) but before the X-ray machine there had been no exciting new technology for the diagnosis or treatment of disease.

The enthusiasm of roentgenologists did not dampen when it was established by the turn of the century that repeated exposure to X rays was injurious to their own health. To the contrary, as Rebecca Herzig observed in the article "In the Name of Science: Suffering, Sacrifice, and the Formation of American Roentgenology," the X-ray pioneers took pride in their painful boils, cancerous lesions, and amputated limbs that were the by-products of their diagnostic work. Frederick H. Baetjer, a roentgenologist at Johns Hopkins, lost eight fingers and an eye to years of working with X rays. "Despite the suffering he has undergone in the interest of science," the *New York Times* reported after the seventy-second operation to save his body, he planned to "continue his work as long as he lives, fingers or no fingers." Elizabeth Fleischmann, famous for her X-ray images of American servicemen wounded in the Spanish-American War, was eulogized as America's Joan of

Arc after she died in 1905 of radiation-induced cancer following a series of amputations.

"The emerging field of roentgenology," Herzig wrote,
"gained definition through the spectacular deaths and mutila-
tions of its adherents." They wore their hideous injuries as
badges of honor. "Scarred and limb-less roentgenologists came
to embody the abstract cause of 'science,' much as stigmata
render palpable the ineffable presence of divinity. At one 1920
professional gathering, historian Bettyann Holtzmann Kevles
reports, so many attendees were missing at least one hand that
when the chicken dinner was served, no one could cut the
meat."

WHEN SANTOS-DUMONT risked his life for aeronautical
progress, he was following the noble, self-sacrificing spirit of
his time, but his motives were not entirely selfless. He enjoyed
being an inventor and an aeronaut but also liked being a showman, and airship trials that courted disaster made for a better
performance. He believed that if his legacy was going to rival
Tiradentes', he needed to do more than perfect the powered
balloon. Men and women had wept at the news of the Brazilian
patriot's gruesome death. As significant an invention as the
flying machine undoubtedly was, Santos-Dumont did not expect people to cry after a successful flight unless they saw the
sacrifices—his brushes with death—that he chose to endure.

In the spring of 1899, Santos-Dumont dismantled *No. 1*,
salvaging the basket, the motor, and the propeller for an airship
that he hoped would better hold its shape. *No. 2* had the same
length as *No. 1* and the same general cylindrical form but was
slightly wider, and, as a result, held 10 percent more gas,
increasing its lifting power by forty-four pounds. He took advantage of the additional carrying capacity by adding a small

rotary fan to supplement the weak air pump, "which," he dryly noted, "had all but killed me." The fan and pump did not force air directly into the belly of the balloon but rather into a separate pocket, a small inner balloon, sewn into the fabric of the outer envelope. That way the air was kept apart from the hydrogen (it is only the mixture of the two, not hydrogen itself, that is highly flammable). The expanding "balloonet" served to prop up the outer balloon envelope so that it kept its cylindrical form.

The first trial was set for May 11, 1899, on the Feast of the Ascension. In the morning, the skies were clear, and Santos-Dumont supervised *No. 2*'s inflation at the captive balloon station in the Jardin d'Acclimatation. "In those days," he recalled, "I had no balloon house of my own. . . . As there was no shed there for me, the work had to be done in the open, and it was done vexatiously, with a hundred delays, surprises, and excuses." By the afternoon, storm clouds blotted the sun and it had started to rain. Because he had no place to store the inflated balloon, he faced an unpalatable choice. He could empty the balloon, wasting the hydrogen and losing the money he had paid for it. Or he could attempt an ascension with an engine that was sputtering from the dampness and a rain-soaked balloon that was heavier, perhaps dangerously so, than it ought to be. He went ahead but as a measure of security tethered the airship to the ground. The drizzle turned into a downpour, and he was unable to rise above the trees before encountering a high-pressure system that compressed the hydrogen so that the balloon visibly shrank. Before the air pump and fan could inflate the balloonet, a strong gust of wind folded up *No. 2* worse than *No. 1* and tossed it into the trees. The balloon ripped, cords snapped, and *No. 2* fell to the ground.

Santos-Dumont's friends rushed over and, finding him in

one piece, strongly admonished him. "This time you have learned your lesson," they said. "You must understand that it is impossible to keep the shape of your cylindrical balloon rigid. You must not again risk your life by taking a petroleum motor up beneath it."

"What has the rigidity of the balloon's form to do with danger from a petroleum motor?" replied Santos-Dumont. "Errors do not count," he continued. "I have learned my lesson, but it is not that lesson." Drenched and a bit scraped up, his panama hat squashed, he was in no hurry to climb out of the dented basket. He surveyed the damage and satisfied himself that the problem was the balloon's long, slender shape, "so seductive from certain points of view, but so dangerous from others." *No. 2*, after just a brief life, would have to be retired, the motor and basket salvaged. In the morning he drew up plans for a plumper airship that would be less sensitive to changes in air pressure.

He made *No. 3* in the shape of a football. "The rounder form of this balloon also made it possible to dispense with the interior air-balloon and its feeding air-pump that had twice refused to work adequately at the critical moment," he wrote. "Should this shorter and thicker balloon need aid to keep its form rigid, I relied on the stiffening effect of a 33-foot bamboo pole fixed lengthwise to the suspension-cords above my head and directly beneath the balloon." Sixty-six feet long by twenty-five-feet wide, *No. 3* had a gas capacity of 17,650 cubic feet, nearly three times that of *No. 2*. When filled with hydrogen, his third airship also had three times the lifting power of his second airship and twice that of the first. The lifting power was more than required, and so he was able to substitute ordinary lamp gas, which was cheap and obtainable everywhere, for the scarce and expensive hydrogen. Although illuminating gas had only half the lifting power of hydrogen, *No. 3* could still carry aloft 50 percent more weight than the hydrogen-filled *No. 2*. In fact *No. 3* could transport a motor, the basket, and the rigging, as well as the aeronaut, with 231 pounds to spare for emergency ballast and a full lunch.

Santos-Dumont scheduled *No. 3*'s first flight for the thirteenth of November, over the protest of skittish members of the newly formed Paris Aéro Club who urged him not to fly on an unlucky day. (France was notorious in its dread of the number thirteen; a *quatorzième*, or professional fourteenth guest, could be hired on the spur of the moment to round out an otherwise ill-fated dinner party.) And November 13, 1899, was not just any unlucky day—it was the day on which centennial alarmists had predicted the world would end. Santos-Dumont enjoyed mocking the superstitions of others. He once rounded off the pay of a triskaidekaphobic housekeeper to a multiple of thirteen and gave her a necklace with thirteen beads. But he had his own peculiar beliefs. "He only entered a place with his right foot first," recalled Antoinette Gastambide, whose father manufactured one of his engines. "He told me that whenever he flew he would wrap a female stocking around his neck," hidden under his shirt so that no one knew. "It was the stocking of Madame Letellier, one of the most famous women in Europe, who had had a lot of luck in her life." Before he ascended, he would also never say good-bye for fear that it would be his last farewell. He could not sleep unless his hat was next to him. As for numbers, he went out of his way to avoid the number fifty, refusing to carry fifty-franc notes or fifty thousand reis bills in his wallet, and later in life—after a scary crash on the eighth of the month—he shunned the number eight. His friends surmised that his preference for flying on "unlucky" days was his way of thumbing his nose at the obvious dangers of aerostation. In general, he preferred to ascend on days of historic importance such as the Fourth of July, Brazil's Independence Day (September 7), or the Feast of the Ascension.

On November 13, 1899, the weather was unremarkable—a cool, crisp day with no signs of precipitation—and the world showed no sign of coming to an end. Santos-Dumont spent the morning inspecting the airship, testing the motor, and checking the all-important exhaust valve. By early afternoon, his workmen had filled the balloon with lamp gas and he was ready to take off from the Parc d'Aérostation in Vaugirard. His friend Antônio Prado asked him whether he was afraid to go up after the two close calls in his previous airships. Santos-Dumont confessed that he was nervous. Prado wanted to know how he faced the fear. "I grow pale," he said, "and try to gain control over myself by thinking of other things. If I do not succeed, I feign courage before those watching me, and face the danger. But even so I am still afraid."

At 3:30 P.M. Santos-Dumont set off on his most successful flight to date. As soon as he was in the air, he headed for the Eiffel Tower. "Around that wonderful landmark for twenty minutes, I had the immense satisfaction of describing circles, figure eights, and whatever other maneuvers it pleased me to undertake, and in all directions, diagonally up and down as well as laterally," he recalled. "I had at last realized my fullest expectations. Very faithfully the airship obeyed the impulse of propeller and steering-rudder." From the Eiffel Tower he made a straight course to the Bois. He did not want to return to Vaugirard because the balloon shed there was surrounded by houses, which meant that there would be little room for error when landing, and the wind that had picked up would make the descent even riskier. "Landing in Paris, in general, is dangerous for any kind of balloon," he said, "amid chimney pots that threaten to pierce its belly and tiles that are always ready to be knocked down on the heads of passersby." So he chose to touch down in the Bois, this time in the most controlled way "at the exact spot where the kite-flying boys had pulled on my guide rope and saved me from a bad shaking up."

Santos-Dumont inspected *No. 3* and was pleased that it had not lost any gas whatsoever. "I could well have housed it overnight," he recalled, if he had had a place to shelter it, "and gone out again in it the next day! I had no longer the slightest doubt of the success of my invention."

That night at Maxim's he boasted about his achievement. Having made one controlled descent, the rooftops no longer seemed so threatening, and he wagered good money that he could land *No. 3* at any specified place in the city. To tweak the members of the Automobile Club, he bragged that he was going to descend in a dirigible on the roof garden of their clubhouse in the place de la Concorde. He told everyone in the restaurant that he was "going into air-ship construction as a sort of life-work."

He contacted the Paris Aéro Club, which had purchased land in Saint-Cloud, just west of the Bois, and persuaded the club to let him build a giant aerodrome, a balloon hangar, at his own expense, complete with a hydrogen-generating plant and a state-of-the-art workshop. He wanted the hundred-foot-long aerodrome to have thirty-six-foot-high doors so that an inflated airship could easily be moved in and out. But again he encountered resistance to his plans. "Even here," said Santos-Dumont, "I had to contend with the conceit and prejudice of the Parisian artisans, who had already given me such trouble at the Jardin d'Acclimatation." They declared that the sliding doors would be too big to open properly. "Follow my directions," he replied, "and do not concern yourself with their practicability. I will answer for the sliding." They were still reluctant. "Although the men had named their own pay," he said, "it was a long time before I could get the better of this vainglorious stubbornness of theirs. When finished, the doors worked—naturally." (Three years later the Prince of Monaco would build him an even bigger aerodrome, and the fifty-foot-high doors that Santos-Dumont requested would have the distinction of being the tallest working doors in the world.)

While the Saint-Cloud aerodrome was under construction, Santos-Dumont continued to fly *No. 3,* which did not require the elaborate preparations that its predecessors did. "To fill five hundred cubic mètres with hydrogen takes all day, whereas with the ordinary burning gas it takes only an hour," he told the *New York Herald.* "Think how much time is saved! I have only to look out my window and see what are the weather indications, and if they prove favorable I am in my balloon an hour afterward." Because he no longer ascended in bad weather, and the airship was demonstrably more stable than its predecessors, the flights were by and large uneventful, until the final one, when the rudder fell off and he had to make an unplanned descent. Luckily there was an open space, the plain at Ivry, below him. He made dozens of trips in *No. 3,* and set a record for the longest time aloft, twenty-three hours.

He would have designed a new rudder for *No. 3* if it were not for a challenge laid down at a meeting of the Paris Aéro Club in April 1900. To stimulate aerostation in the new century, Henry Deutsch de la Meurthe, a petroleum magnate and founding member of the club, announced that he was offering a hundred thousand francs (twenty thousand dollars) to the first airship that "between May 1 and October 1, 1900, 1901,

1902, 1903, or 1904 should rise from the Parc d'Aérostation of the Aéro Club at St. Cloud and, without touching ground, and by its own self-contained means on board alone, describe a closed curve in such a way that the axis of the Eiffel Tower should be within the interior of the circuit, and return to the point of departure in the maximum time of half an hour. Should more than one accomplish the task in the same year, the one hundred thousand francs were to be divided in proportion to their respective times." Deutsch added that if the prize was not claimed in any given year, he would, as a gesture of encouragement, award the interest on the hundred thousand francs to the aeronaut who had accomplished the most in the previous twelve months. Santos-Dumont, who had attended the Aéro Club meeting, told his friends that Deutsch would not have to part with the interest. He planned to come away with the prize, and the attendant glory, before the year was out.

The Eiffel Tower was three and a half miles from Saint-Cloud, and so an airship would have to travel fourteen miles an hour to make the round-trip in thirty minutes (actually the necessary speed was probably closer to fifteen and a half miles per hour if one considers the time lost in turning around the tower). *No. 3* had attained a speed of only twelve miles an hour, although Santos-Dumont calculated that in calm air, with the motor and propeller working perfectly, it might reach twenty. But he knew, of course, that he could not count on ideal weather, and in anything less the "balloon was too clumsy in form and its motor was too weak." He needed a faster airship, and he set to work on *No. 4* at once.

The prize was sufficient motivation, but he was also inspired by the upcoming Paris Exposition designed to kick off the new century. Many of the world's top aeronautical experts were scheduled to gather there for the International Aeronautical Congress and debate the future of flying machines. Santos-Dumont had little patience with theoretical discussion. He wanted those who questioned the dirigibility of balloons to witness him flying *No. 4* high above them.

On July 10, 1900, the press reported that he would soon test his new airship. According to the *New York Times,* "friends at the Automobile Club joke daily with M. Santos-Dumont about the contemplated trial of the new balloon. . . . They tell him he will be killed, but the aeronaut's faith in his system is unshaken and he is determined to try it." On August 1, he unveiled *No. 4* to his fellow aeronauts. They were stunned that he had dispensed with the protective wicker basket and appeared to "ride on a stick like a witch," exposed in the open air atop a bicycle seat. He had dispensed with the basket, he told them, because it was too heavy and a needless luxury. He had managed to salvage the thirty-three-foot bamboo pole from *No. 3* and rather than putting it above his head to give the balloon envelope rigidity, as he had in *No. 3,* he now in essence sat on it. Bolstered by a spiderweb of vertical and horizontal cross pieces and taut cords, the pole also supported the motor, the petroleum fuel tank, and the propeller.

"As Dumont sat straddling the bicycle saddle, his feet on the pedals, one hand on the brass cylinder containing his supply of petroleum essence and the other on the larger brass cylinder containing his supply of water ballast, I could not help but wonder at his courage in trusting himself to the air so unprotected," Sterling Heilig wrote in the *Washington Star.* " 'If you should faint while up there you would fall and be dashed to pieces,' I exclaimed. 'I shall not faint,' he answered. 'I have no fear of falling out of the framework,' repeats Santos-Dumont, 'because I know that I shall keep my head. A basket would be a good thing for anyone proposing to lie down in the bottom of it and shut his eyes, but I must be able to control all this machinery, and must be situated so that I can do it properly. As it is, I have everything within reach of my hands and feet. I ask no better protection.' "

The bicycle saddle, he assured his dumbfounded colleagues, was comfortable and the whole bicycle frame quite functional. The handlebars, for example, controlled the rudder, and the pedals started a two-cylinder seven-horsepower engine, an improvement over the three-and-a-half-horsepower tricycle motor. By means of a long shaft, the engine drove a thirteen-foot-wide propeller consisting of two large wings of fabric stretched over a steel frame. In *No. 3* the propeller was mounted in the rear, where it served to push the craft through the air. In *No. 4* it was positioned at the front of the bamboo pole, where, at one hundred revolutions per minute, it pulled the airship. Though the huge hexagonal rudder—silk stretched over wooden rods—occupied seventy-five square feet, it was light enough to be attached directly to the balloon envelope.

Each rope in the tangle around the bicycle seat had a specific purpose. They controlled the shifting of the weights (the guide rope and bags of sand), the striking of the engine's electric spark, the opening and shutting of the balloon's valves, and the turning of spigots to release water ballast. One cord even served to rip open the balloon in the event of an emergency. "It may easily be gathered from this enumeration," said Santos-Dumont, "that an airship, even as simple as my own, is a very complex organism; and the work incumbent on the aeronaut is no sinecure."

Those who examined *No. 4* were worried that he might have too much to do. To them, the machine, though "marvellously ingenious," seemed too complex to operate. "It so bristles with apparatus of all kinds that it seems to me that no one man could work it all," London's *Daily Graphic* reported. "It is perfectly possible that at a given moment M. Santos-Dumont may desire to throw his propeller out of gear, stop his engine, and port his helm." It was not just a question of remembering, in the heat of the moment, which cord did what, but of pulling them properly all at once. "As things stand, being alone, he could not do so."

The balloon itself was less controversial, though no less striking, than the rest of the airship. It looked like "a huge yellow caterpillar," the *Herald* said. "Some idea of its proportions can be gathered from the size of the shed in which it is housed. . . . If a traveler were to come across this big shed he would be puzzled as to whether it was an embryo church or a diminutive grain elevator." Santos-Dumont viewed the balloon as a compromise, in form and capacity, between *No. 3* and its predecessors. With a gas-capacity of 14,800 cubic feet, it was ninety-five feet long and seventeen feet at its greatest diameter, but no longer a cylinder terminated by two cones. "It was, rather, elliptical in form," said Santos-Dumont, "and while not a return to the slender straightness of *No. 1,* it had so little of *No. 3*'s pudgy compactness that I thought it prudent to put the compensating air-balloon inside it again, this time fed by a rotary ventilator of aluminum. Being smaller than *No. 3,* it would have less lifting power; but this I made up by going back to hydrogen gas," which he could now generate himself at his new plant. *No. 4* may have been smaller in volume than *No. 3,* but at ninety-five feet, it was his longest and most impressive airship to date.

For two weeks in August, he took *No. 4* up almost daily. The balloon barely leaked, so he would just park it in his new aerodrome between flights. It was as easy to maintain as a car, he said, and a lot more fun. Parisians turned out to watch him walk the giant gasbag out of its garage and mount the precariously protruding bicycle seat. The engine was directly in front of the seat, and it splattered sparks, ashes, and oil onto his pristine suits, which he sent out daily to be cleaned. Some of the aeronauts who arrived early for the exposition joined the crowd at Santos-Dumont's aerodrome. Others caught a distant glimpse of him soaring about.

On September 19, the International Aeronautical Congress adjourned the official proceedings to witness a demonstration. Santos-Dumont planned to take *No. 4* around the Eiffel Tower, but a stiff breeze made him reconsider and train his sights on a simple "spin round the Bois." Even so, as London's *Daily Express* observed, "the voyage which M. Dumont is about to undertake is uncommonly dangerous. There appears to be here all the elements of a catastrophe. A red-hot motor within a yard or so of some thousands of cubic feet of hydrogen is enough to give the most stout-hearted the 'creeps,' but M. Dumont makes light of it." By 3:30 P.M. several hundred people had gathered. The winds had picked up, and the normally unflappable Santos-Dumont was nervous about taking the balloon out of the shed. Because some of his esteemed guests had crossed oceans to meet him, he felt compelled to put on a limited demonstration. To cheers and appreciative laughter, he led the balloon outside, but the crowd grew silent when a strong gust battered the airship against the building. The rudder snapped. "It would have taken two hours to splice the defects and put the doubtful steering gear in its original form again," the *Herald* reported, "and so it was not attempted." Santos-Dumont announced that it would now be foolhardy for him to attempt a free ascent, but that he would go up a short distance with the airship tethered to the starting blocks to show his guests that the motor was indeed a match for the stormy weather. At first he started with the wind and climbed quickly to an altitude of about seventy feet. In response to shouts from the crowd, he turned the airship around and began to move against the ten-mile-an-hour wind. "This was the crucial test which everyone wanted to see!" the *Herald* said. "In the teeth of the wind," the airship was gaining ground at about four miles an hour, but the course was hard to maintain because without a rudder *No. 4* tended to slew around until it was broadside. Even the limited demonstration was impressive. "Skeptics were convinced, and everyone was unanimous that M. Santos-Dumont could propel his balloon by his motor," the *Herald* concluded. But with the rudder out of commission, some questioned whether he had really "solved the problem of steering his airship."

•

SAMUEL PIERPONT LANGLEY, one of the attendees at the
Aeronautical Congress, returned for a private showing during
the last week of September 1900. Langley was the head of the
Smithsonian Institution and arguably the world's foremost authority on heavier-than-air flight, the effort to build gliders and
winged flying machines. While airships were largely the domain of Europeans, specifically Frenchmen (and a frenchified
Brazilian), the development of the airplane was more of a
worldwide phenomenon, with the United States taking the
lead.

Judging from the story of Icarus, and similar myths in Asian
and African cultures, the fascination with manned flight was
universal and went back to antiquity. There were dozens of
legends about winged men, but there were no myths about
people ascending to the heavens in balloonlike contraptions.
The disparity can be explained by the examples offered by the
natural world. With the exception of the occasional effervescent bubble, there were few instances of airborne orbs but an
abundance of heavier-than-air birds flapping their way across

the sky. The first aeronauts consequently modeled their flying machines on birds, not bubbles.

About the year 1500, Leonardo da Vinci penned the earliest documented proposal on how to fly. He wrote thirty-five thousand words on flying machines and drew five hundred sketches of men with artificial wings. But Leonardo flew only on paper. Other men in the Middle Ages and Early Renaissance, like the Italian mathematician Giovanni Battista Danti, also called Daedalus of Perugia, glued feathers to their arms and leapt from towers, only to crash and break their limbs. In 1660, a French tightrope dancer named Allard boasted that he would fly from the terrace of Saint-Germain to the woods of Vesinet if the king consented to watch. Louis XIV promptly agreed, and a now-reluctant Allard jumped from the tower and smashed his skull on the stone patio.

The early aeronauts, in their attempts to mimic bird flight, made the mistake of concentrating on the flapping of wings. In the eighteenth century, naturalists convincingly demonstrated that man did not have the muscle strength to flap wings that were large enough to propel him through the air, and technologists had little success in building a machine, called an ornithopter, that would do the flapping for them. Progress in heavier-than-air flight came only in the early nineteenth century when inventors abandoned flapping and based their machines on the other form of locomotion that birds did so well—gliding with wings held relatively stationary.

The seminal figure was Sir George Cayley, an English engineer, who had gone into aeronautics because of accounts of the Montgolfiers' balloon flights, which had excited him as a ten-year-old boy. Because other investigators were successfully extending the Montgolfiers' work, Cayley decided to pursue the less-crowded field of heavier-than-air aeronautics. He constructed model gliders in 1809 and full-size ones by the 1850s. Engineers like Otto Lilienthal and Octave Chanute built on Cayley's efforts to the point where, at the beginning of the twentieth century, glider design was the most promising area of aeronautical research.

In pursuing lighter-than-air flight, Santos-Dumont was a maverick among contemporary aeronauts. The airship had the obvious advantage over the plane that if the propeller or motor failed, the buoyant gasbag would prevent it from falling out of the sky. But the sheer bulk and ungainliness of the gasbag seemed to limit an airship's speed—a detraction in a velocity-conscious age. Every week it seemed bicycles, steamships, trains, and automobiles were setting new speed records. Most aeronauts wanted to build the fastest-possible flying machine and planes, not balloons, seemed to be the answer even if they had not yet left the ground.

LANGLEY WAS twice Santos-Dumont's age, but the two men got along at once. In the Brazilian's crisp suits, even if they were a bit foppish, the Boston Brahmin, one of whose forebears was the president of Harvard and the first American to write a book on astronomy, recognized a fellow man of fine manners. On more than one occasion Langley had reprimanded a Smithsonian employee for wearing a loose tie or slumping in a desk chair. Before he hired someone for an aeronautical project, he asked for references so that he could determine whether the candidate was of good moral character; anyone could borrow nice clothes for an interview, but that did not mean he was really a gentleman. This bear of a man abhorred profanity, and demanded of his mechanics that they speak refined English even if one of their cherished new motors or test crafts was on the verge of exploding.

Langley had taken up aeronautics after an illustrious career as an astronomer. He became interested in the stars as a child in Roxbury, Massachusetts, where he looked through his father's telescope not just at the heavens but at the construction of the Bunker Hill Monument. In the late 1870s Langley invented a new instrument, the bolometer, to measure heat from the sun. From data collected in 1881 atop Mount Whitney, in southern California, he discovered that the solar radiation had a much broader spectrum than was previously thought, and he did pioneering work on the extent to which the sun's energy was absorbed by the earth's atmosphere. His discoveries earned him honorary degrees at universities around the world and membership in the National Academy of Sciences in Washington, the Royal Society in London, and the Accademia dei Lincei in Rome. There were few academic honors that escaped him. He was president of the American Association for the Advancement of Science and was the natural choice to head the Smithsonian when its former director died on the job in 1887. He achieved all this without completing college, a fact that was not lost on Santos-Dumont.

Langley took up heavier-than-air flight in earnest the year before he joined the Smithsonian. In the mid 1880s, American scientists by and large viewed aeronautics as the quixotic pastime of backyard tinkerers and medieval tower-jumpers. In 1886, Chanute, a railway engineer who experimented with gliders and was a one-man clearinghouse for information on what anyone in America was doing in aeronautics, believed that true progress would come only if more engineers and scientists could be attracted to the effort, and this would happen only if the flaky image of the field itself was changed. As Smithsonian historian Tom Crouch explained in *A Dream of Wings,* Chanute did not want to risk his own reputation by calling outright for more research but he found a way of quietly sounding out his colleagues. Chanute was in charge of setting the program for the Mechanical Engineering section of the 1886 Buffalo meeting of the American Association for the Advancement of Science, and he worked behind the scenes to schedule two lectures on aeronautics by Israel Lancaster, an amateur ornithologist who had built hundreds of so-called soaring effigies, imitation birds designed to show that machines modeled after birds could actually fly.

Chanute gave Lancaster's first session the innocuous title "Soaring Bird," but it stood out in a program largely devoted to more traditional concerns of mechanical engineers. Lancaster told his learned audience that his effigies had soared five hundred yards and stayed up more than fifteen minutes. He then proudly explained his vision for an effigy with a two-hundred-square-foot wingspan that would carry aloft a man. Chanute's fears came true; the ornithologist's listeners "unanimously joined in reviling and laughing at him," the *Buffalo Courier* reported. Now the word was out at the normally staid conference that Lancaster's second session was the place to go for comic relief. He was scheduled to demonstrate the effigies but, with the room overflowing with chuckling scientists, he chickened out. The crowd turned on him, hooting and hollering. One heckler offered a hundred dollars if anyone could make an effigy fly; another upped the ante to a thousand dollars. Chanute, perhaps expecting the worst, was not even in attendance this time. But Langley happened to be there, and he did not join in the laughter. He sat there calmly, in the middle of the circus, intrigued by the promise of human flight.

"How is it that a turkey-buzzard is able to sustain itself in the air for hours at a time, apparently without the slightest motion of its wings?" Langley wondered. "These birds weigh five or ten pounds, they are far heavier than the air they displace, they are absolutely heavier than so many flat-irons. If we saw cannon-balls floating through the air like soap bubbles we should look upon it as a sufficiently surprising matter, if not as a miracle. The only reason we are not surprised at the soaring bird is that we have seen it from childhood. Perhaps if we had seen cannon-balls floating in the air from our childhood we would not stop to inquire how they did it, any more than we now stop to inquire how the turkey-buzzard soars without flapping its wings."

Langley had been working for twenty years as the director of the Allegheny Observatory in Pittsburgh and was finishing his service in order to join the Smithsonian. Even as he shuttled between Pittsburgh and Washington, he lost no time in starting work on the problem of flight. He believed that model-building was not enough to conquer the air. As an astrophysicist, he knew that aeronautics was not yet a science. Most of his fellow physicists had a harsh view that it would never be a science because Newtonian principles seemed to rule out the possibility of manned flight altogether. These physicists argued that there was a paradoxical relationship between wingspan and wind resistance. Wings large enough to transport a man would generate substantial wind resistance that could be overcome only by a very powerful engine. But such a powerful and presumably heavy engine would require even larger wings to stay aloft, which in turn would lead to greater wind resistance that would require an even more powerful engine. That, of course, would mean still bigger wings, and so on ad infinitum.

Langley hoped to prove the Newtonians wrong, but, with Lancaster's humiliation fresh in his mind, he was quiet about his intentions. Langley thought the issue could not be settled without extensive empirical data on actual wind resistance. To that end, at Allegheny he constructed a huge "whirling table"— a primitive wind tunnel—two thirty-foot horizontal arms pivoting on a vertical shaft turned by a steam engine. On the ends of the arms, which attained speeds of seventy miles an hour, he "twirled" a dead albatross, a condor, a buzzard, and various artificial wing shapes to see how they performed. He did all this behind closed doors, spiriting the stuffed birds in and out, and cryptically referring to the experiments as work in pneumatics. His chief finding, the so-called Langley Law, was encouraging to the disciples of Icarus: As the speed of an object increased, less power—not more power—was required to sustain its flight. He began constructing his own model planes, starting with rubber-band-powered "toys" and gradually increasing their size. In 1891, he published his conclusion that heavier-than-air flight was not only possible but could be done with existing engines. But Langley's colleagues snickered when he predicted that a twenty-pound, one-horsepower steam engine could propel a two-hundred-pound airplane at a speed of forty-five miles an hour.

In 1894, Langley attended a meeting in Oxford of the British Association for the Advancement of Science in which the validity of the Langley Law was fiercely debated. The British scientists were not predisposed to accept his conclusions. Langley's chief antagonist was William Thompson Kelvin, the dean of British science, who had become a professor of physics at the University of Glasgow in 1846 at the age of twenty-two, having published his first paper when he was sixteen. Before the Oxford meeting, Lord Kelvin had gone on record as flatly dismissing the possibility of large flying machines. At the meeting, he accused Langley of making unforgivable computational errors. The dean of American science defended himself firmly but respectfully; Langley could have reminded Kelvin that when the good lord had wandered too far afield of his expertise in thermodynamics, he had been famously wrong. (Kelvin had made the mistake of challenging Charles Darwin in the late 1860s.) Another member of the British peerage, John William Strutt Rayleigh, the discoverer of argon and an optics pioneer whose work explained why the sky was blue, was kinder. While not exactly rushing to Langley's defense, he encouraged him to defuse his critics by demonstrating mechanical flight. To salvage his reputation, Langley now had little choice.

To solve the problem of mechanical flight, Langley had been availing himself of the Smithsonian's substantial resources: the machine shops, the wide variety of experts, the ample funding. He constructed six large model planes, with tandem wings, one behind the other, like a dragonfly. Each set of wings jutted up like a squashed letter V. The biggest model had a thirty-foot wingspan and weighed thirty pounds. He called the planes Aerodromes—a confusing name because that was what balloon sheds were often called. Powered by steam engines positioned between the tandem wings, the Aerodromes were launched by a twenty-foot catapult mounted on a houseboat in the Potomac River. Langley had tested the giant catapult at the National Zoo. Bystanders who got a glimpse of the strange-looking machine may have wondered whether it was a device of last resort for subduing large angry mammals.

On May 6, 1896, accompanied by his friend Alexander Graham Bell, Langley traveled to Chopawamsic Island, thirty miles south of Washington. There the Potomac was sufficiently wide for a test flight. It was also isolated enough that Langley could not do any damage to innocent parties, and it seemed safe from the prying eyes of those who might lambaste him. At 1:10 P.M. Aerodrome *No. 6,* so long in construction, got caught on the catapult and the left wing snapped before Langley's assistant could launch it. But the model's understudy, *No. 5,* turned in a stunning debut. At 3:05 P.M. the fifth Aerodrome, whose front wings spanned fourteen feet, climbed perhaps a hundred feet at twenty to twenty-five miles an hour and, even to Langley's astonishment, traveled more than half a mile. The flight path was a graceful curve, the craft falling into the river only when its steam engine ran out of water. "Its motion was so steady," reported Bell, "that I think a glass of water on its surface would have remained unspilled." Six months later, the repaired *No. 6* set a new record of four-fifths of a mile, at a speed of thirty miles an hour. Langley had demonstrated for the first time in history that a large heavier-than-air machine could fly under its own power—and the inventor of the telephone was there to bear witness. Even Lord Kelvin took notice, although he still maintained, but not quite as firmly, that manned flight was still impossible.

After his success in 1896, Langley had thought of quitting aeronautics and letting a new generation of scientists extend his work. His doctor had warned him that continued devotion, with the attendant worries and stress, would shorten his life. But any plans the Smithsonian director had of retiring were derailed when President William McKinley called on him. McKinley was preparing for the Spanish-American War, and he wanted Langley to build a plane that could survey the enemy and, better yet, carry missiles. Langley could not resist the president's call to service. The president dispatched a young Theodore Roosevelt, assistant secretary of the navy, to persuade Congress to appropriate fifty thousand dollars for Langley's continued work. Langley accepted the funding only after protracted negotiations in which he won assurance that there would be no oversight, military or otherwise, of how he spent the money. He also persuaded Congress to keep his work and the funding secret, on the grounds of national security, but word got out anyway. The real reason he wanted to keep his experiments clandestine was fear that the Lord Kelvins of the world would relish any failure.

Santos-Dumont, though better funded than most aeronauts, was nonetheless envious of Langley's resources but sympathized with his struggle to protect his work from meddlers and doubters. Despite all his accomplishments, Langley was at heart a shy bachelor whose reticence could come across as haughty aloofness. And he had a few quirks that engendered animosity; for instance, he forbade his employees to walk down the hall in front of him.

Langley certainly had his detractors, not just former employees wounded by his imperiousness but distinguished men of science, who were wishing not so quietly for his downfall. His critics would get their wish in three years, and in a most spectacular way, but at the time Langley came to France, at the dawn of the new century, his achievements in aeronautics were unprecedented. The scaling up of his Aerodromes into full-size planes turned out not to be a simple task, and President McKinley was never able to enlist a plane in his fight with Spain. In the four years since Langley demonstrated mechanical flight, he had come to realize that the steam engine was not up to the task. To make a suitable motor for the first man-carrying plane, he had commissioned Stephen Marius Balzer, an ex-watchmaker from Tiffany & Co. who had built New York City's first homegrown automobile in 1894.

Balzer designed an elegant five-cylinder rotary engine—the entire engine spun around the driveshaft. The rotary engine had the advantage of cooling itself as it spun in the air, eliminating the need for a separate water-cooling system, but it had a serious drawback: inadequate lubrication. It was not possible to keep the engine oiled because centrifugal force pushed the lubricant to the outer ends of the cylinders. Displeased with Balzer's work, Langley had come to Europe in the summer of 1900 chiefly to investigate the state of European automobile engines. Although Santos-Dumont's engine was too weak for Langley's needs, the American admired the strides he had made in lighter-than-air flight.

Santos-Dumont welcomed Langley's attention. The Smithsonian director's interest gave him a legitimacy in the scientific world that he had not enjoyed before. The two men talked long into the night about the future of flight. They found that they shared the same stubborn temperament. Neither took criticism easily, and both insisted that workmen follow their respective instructions to the letter and keep any disagreements to themselves. Each man had shown that he could stick to his convictions in the face of experts who held contrary views.

Santos-Dumont had not given airplanes much thought before he met Langley, but the Smithsonian director's enthusiasm and optimism were infectious. Even before Langley had a few drinks in him, he spoke of his vision of giant winged machines whizzing people around the globe. Santos-Dumont promised him that as soon as he won the Deutsch prize, he too would work on heavier-than-air flight.

But for now there was the airship to perfect. He recognized that *No. 4*, which Langley had so kindly praised, was not reliable enough to make it around the Eiffel Tower in thirty minutes. Perhaps it was for the best, he thought, that he was not able to demonstrate an untethered *No. 4* to the Aeronautical Congress. Its slow speed may have disappointed the experts and caused them to dismiss aerostation prematurely. Although *No. 4*'s engine was twice as powerful as *No. 3*'s, the airship itself was heavier, and so there was not much gain in speed. Santos-Dumont knew he needed more power, so he doubled the number of cylinders in the engine. To compensate for the increased weight of the four-cylinder motor, he enlarged the balloon. He cut the silk envelope in half and added a piece of cloth, "as one puts a leaf in an extension-table." But the 108-foot-balloon was now eight feet too long to fit in the aerodrome, and so he ordered his workmen to rip off the back wall of the hangar and extend the length. The work was completed in fifteen days, and with the Aeronautical Congress still in session, he was eager to make another flight.

But the weather did not cooperate. Fierce rains, so typical of Paris in autumn, ruled out the ascent. "After waiting with the balloon filled with hydrogen through two weeks of the worst possible weather," Santos-Dumont recalled, "I let out the gas and began experimenting with the motor and propeller." After much tinkering, he succeeded in boosting the propeller's speed 50 percent, so that it made 140 revolutions per minute. "The propeller turned with such force," he recalled, "that I contracted a severe cold in its current of cold air." The respiratory ailment turned into pneumonia, and at the end of October 1900 he was forced to call off further trials. To recover his strength, he headed to the French Riviera for the duration of the fall and winter. He also hoped that the more propitious weather in Nice would enable him to get in a few balloon trials before spring.

[CHAPTER 6]

AN AFTERNOON IN

THE ROTHSCHILDS'

CHESTNUT TREE

PARIS, 1901

•

SANTOS-DUMONT NEVER actually tested the four-cylinder
engine in *No. 4,* on the Riviera or elsewhere. He changed his
mind about the airship's stability. Once he had prided himself
on his resourcefulness in salvaging the bamboo pole from
No. 3 and using it as the "platform" for *No. 4,* but now he
conceded that his critics were right in denouncing it as dangerously flimsy. When his health was restored, he rented a
small carpentry shop in Nice and built himself his first true
airship keel, a sixty-foot-long narrow pine lattice, triangular in
form, that was at once rigid and light at ninety pounds. One
morning, while pacing in the workshop, studying the keel, he
tripped over a coil of piano wire. Cursing the wire, he was
nonetheless impressed by its strength. After ascertaining that
it was destined for the trash bin, he decided to use it to reinforce the pine lattice.

"Then followed what turned out to be an entirely new idea
in aeronautics," he recalled. "I asked myself why I should not
use this same piano-wire for all my dirigible balloon suspensions in place of the cords and ropes used in all kinds of
balloons up to this time. I did it; and the innovation turned

out to be peculiarly valuable. These piano-wires, 0.032 inch in diameter, possess a high coefficient of rupture and a surface so slight that their substitution for the ordinary cord suspensions constitutes a greater progress than many a more showy device. Indeed it has been calculated that the cord suspensions offered almost as much resistance to the air as did the balloon itself!"

Now that he had constructed a relatively stable platform for himself, the engine, and the propeller, he rethought where each of them should be situated. It had occurred to him, when he sat behind the propeller in *No. 4*, that the guide rope could cross the propeller and be diced to pieces. In *No. 5* he wanted the guide rope and control wires to be as far from the rotating blades as possible, so he planned to put his bicycle seat at the bow and return the propeller to the stern. In *No. 4* he had sat next to the motor, so that he could closely monitor it. But the motor was deafening, not to mention filthy. Engine technology had improved in the past few months, as more and more automobiles hit the roads, and so there was less need to watch the motor constantly. In the interest of saving his hearing and reducing his cleaning bills, he decided to move the engine to a position well behind him at the center of the keel. So that he would feel less exposed, he also planned to restore the narrow basket. He returned to Paris early in 1901. When he crossed the city's limits, zealous customs officials, who were responsible for levying taxes on goods brought in from the provinces, did not know what to make of the sixty-foot keel. Here was a man with Brazilian papers claiming to be transporting the backbone of an airship. There was no aeronautics category on their list of taxable goods, and so the stymied bureaucrats confiscated the keel and pondered what to do. Santos-Dumont feared that they would damage it, but they returned it unharmed a week later, calling it a piece of fine cabinetry and imposing the maximum levy.

The customs officials' ignorance was understandable. Santos-Dumont was set on mastering the skies at a time when the vast majority of Europeans and Americans had not yet traveled along the ground in an automobile. Even heads of state, who had access to cars if they wanted them, were nervous about driving. On July 12, 1901, President McKinley summoned the courage to take his first automobile trip, in his hometown of Canton, Ohio. "Hitherto he had avoided horseless carriages in Washington, Canton, and elsewhere," the press reported. But when his friend Zeb Davis drove up while the president was taking a constitutional, McKinley put his fears behind him and accepted a ride. "A run was taken about the city. The President seemed to enjoy the trip very much. He, however, tightened his grip on the seat and stiffened his backbone as the sharp curves were made about the turns in the street. Davis let out his machine at several places, and the President seemed to be pleased with the run, although he preferred straightaway going. During the run a bicycle rider had a close call, but the skillful steering of Davis saved the wheelman from injury."

Two weeks later, King Alfonso XIII of Spain did not fare much better. At his seaside palace in San Sebastian, he took his first car ride, accompanied by his mother. "The carriage swept suddenly out of the grounds of Miramar palace," the *Herald* reported, "and . . . threw the sentinels into a panic. The alarm was sounded and the palace guard turned out. The royal household was in a tumult, but the King restored order, personally appearing and offering his assurances that anarchy had not broken loose."

At the turn of the century, officials on both sides of the

Atlantic were confronting questions about the proper use of cars. On November 13, 1899, George Clausen, the head of the New York City Parks Department, drove the first automobile into Central Park. Cars had been prohibited out of concern that they would startle the carriage horses. New York's new automobile club was pushing Clausen to lift the ban, and he wanted to see for himself how the animals would react. The horses initially bolted, but they eventually got used to his car. That was encouraging, but Clausen, who was new behind the wheel, had another concern. Were cars really safe enough to allow on the park's congested streets? "To drive a spirited horse through the Park without accident requires skill," he said. "It will require even more of a different kind of skill to handle an automobile on the drives. The greatest skill is required in taking a machine up or down a crowded hill. The difficulty lies in properly regulating the speed. If the driver of a horse is obliged to suddenly check his pace, and he is close pressed at the rear, he can raise his whip as a warning signal to those behind him. The driver of an automobile cannot give such warning, for hands, arms, and feet are constantly engaged with the various levers and buttons."

Newspapers in Paris, London, and New York were sending reporters to each trial of Santos-Dumont's airships, but the French military showed no interest. It did not want to be bothered with flying machines at a time when it was still trying to determine the role of automobiles in warfare. French army commanders had been testing horseless carriages in military exercises, and in October 1900, they reported to the minister of war that they were unanimously in favor of using them on the battlefield. According to the report, automobiles had proved effective in enabling army corps commanders to go from point to point in the fighting line to see for themselves the disposition of the troops, instead of relying on verbal or written reports, and that they are of great use for conveying staff officers and orderlies quickly to the various parts of the battlefield. On the other hand, all attempts to utilize automobiles on the fighting line or in aid of outpost or reconnoitering work failed, because a single bullet in the mechanism would render the machine useless, as an automobile is an easy mark, presenting a large, vulnerable surface and moving with great regularity.

SANTOS-DUMONT was determined to win the Deutsch prize in 1901, although he objected to many of the rules, particularly the thirty-minute time limit. Because no one had yet made a clean trip around the Eiffel Tower in any amount of time, there was no need, he believed, to set the bar so high. Club officials thought that his objection stemmed from his fear of failing. Everyone knew that if he could take his time, he could mosey his way around the tower in his existing airship. What was the sense of establishing a prize if it was clear from the start who the winner was? The men of fine breeding at the Aéro Club considered it déclassé to object to the terms of the prize. After all, it was Deutsch's own money and he was entitled to part with it any way he chose.

Santos-Dumont had other objections too. "Another condition formulated by the Scientific Commission [the Aéro Club's aerostation committee] was that its members—who were to be judges of all trials—must be notified twenty-four hours in advance of each attempt," he wrote.

Naturally the operation of such a condition would be to nullify as much as possible all minute time calculations based either on a given rate of speed through perfect calm or such air-current as might be prevailing twenty-four hours previous to the hour of trial. Though Paris is situated in a basin, surrounded on all sides by hills, its air-currents are peculiarly variable; and brusk meteorological changes are extremely common.

I foresaw, also, that when a competitor had once committed the formal act of assembling a scientific commission on the slope of the river Seine so far away from Paris as Saint Cloud, he would be under a kind of moral pressure to go on with his trial, no matter how the air-currents might have increased, and no matter in what kind of weather—wet, dry, or simply humid—he might find himself. . . .

And, finally, I saw that the competitor would be barred by common courtesy from convoking the commission at the very hour most favorable for dirigible-balloon experiments over Paris—the calm of the dawn. The duelist may call out his friends at that sacred hour, but not the air-ship captain!

The Aéro Club wanted to put the dispute about the rules behind it. As a peace offering, early in 1901 the club awarded Santos-Dumont the Encouragement Prize, the four-thousand-franc interest on the Deutsch prize, for his aeronautical accomplishments during the exposition. But he did not accept the overture. He returned the four thousand francs to the Aéro Club with a note asking that the money be used to fund a new prize that had none of the objectionable conditions. "The Santos-Dumont Prize," the note said, "shall be awarded to the aëronaut, a member of the Paris Aéro Club and not the founder of the prize, who, between May 1 and October 1, 1901, starting from the Parc d'Aérostation of Saint Cloud, shall turn around the Eiffel Tower and come back to the starting-point, at the end of whatever time, without having touched ground and by his self-contained means on board alone. If the Santos-Dumont Prize is not won in 1901, it shall remain open the following year, always from May 1 to October 1, and so on until it be won."

Henry Deutsch was distressed by Santos-Dumont's insolence, but there was not much he could do. To the elders of the Aéro Club, Santos-Dumont was a brilliant but impossible nephew whom you had to invite to Thanksgiving dinner not only because he was a family member but also because he owned the turkey farm. Santos-Dumont was the only Aéro Club member who was building machines that captured the world's attention, so it would have appeared petty if the club did not honor his wishes. Deutsch made the best of it by offering a motion at the next meeting that the club bestow its highest reward, a gold medal, on the winner of the Santos-Dumont prize. The Brazilian in turn commended Deutsch for recognizing the importance of the new prize.

Santos-Dumont tired of club politics and buried himself in the task of finishing *No. 5*. He could no longer turn to Machuron to sew the silk envelope. The balloon maker, only a year older than Santos-Dumont, had died in March at the age of twenty-nine after a protracted illness. To commemorate the man who had introduced him to ballooning, Santos-Dumont vowed to make an assault on the Deutsch prize that summer, but first he wanted to show that he could fulfill the terms of his own prize, even if he had prohibited himself from winning it. For two nights he slept in the balloon shed at Saint-Cloud waiting for favorable weather. On the first morning it rained heavily. On the second, Friday, July 12, the skies were calm, and at 3:00 A.M., he took *No. 5* up for the first time. He circled Longchamp five times at low altitude, prompting the night watchman to object and call his superior. (The watchman ended up losing his job for interfering with "history in the making," and Santos-Dumont received an apology.) From Longchamp Santos-Dumont crossed the Bois de Boulogne and made his way toward the Eiffel Tower. One of the cords controlling the rudder snapped, and the airship narrowly missed the tower. Santos-Dumont made an emergency landing in the Trocadéro Gardens and quickly repaired the rudder with the aid of a twenty-foot ladder lent to him by two workmen. He ascended again and properly circled the tower before returning to Saint-Cloud.

He had been gone an hour and six minutes. Waiting for him was Emmanuel Aimé, the secretary of the Aéro Club and one of the few members who was firmly in Santos-Dumont's camp. Aimé, a professor of mathematics, proclaimed the flight a "marvelous, amazing, wonderful success." So did the *Herald*. Dismissing the snapped cord as a trifling accident, the newspaper declared, "There is scarcely any room for the shadow of doubt that M. Santos-Dumont has definitely solved the problem of aerial navigation. . . . The accident only showed unmistakably the practical utility of M. Santos-Dumont's astounding invention. The air-ship came to earth as easily and lightly as a bird and when the defective steering rope connected with the rudder had been repaired, it ascended again, made a complete turn upon itself and returned to its starting point."

Santos-Dumont, emboldened by his new fame (and not at all chastened by the failure to meet the terms of his own prize), was no longer shy about convoking the aerostation committee at the duelist's hour. He announced his intention of winning the Deutsch prize on July 13, a day on which he could once again thumb his nose at superstition. The commission members would have preferred to stay in bed and rest up for the late-night activities that would commemorate the Fall of the Bastille. But at 6:30 A.M. they dutifully arrived at Saint-Cloud, tired and sweaty. "Paris has been trying to rival New York in the matter of a hot spell," the *Herald* gleefully reported, "and, as the absence of electric fans and soda water fountains handicaps the Villa Lumière, this side is keeping its end up pretty well as regards heat strokes, which daily number hundreds, about a dozen of them being fatal." Even the privileged could not escape the heat: The Belgian queen, Marie Henriette, passed out while playing croquet. With the temperature reaching 92 degrees Fahrenheit in the shade, and the air barely cooling down at night, the French military called off training exercises, and the crime and suicide rates soared. "One man smashed his wife's head against a wall," the *Herald* said, "and another threw his better half into the Seine because dinner was late."

Santos-Dumont did not look like he was overheated. He greeted the commission wearing a starched suit with not a bead of sweat on his face. As potential witnesses to history, the commission members were also wearing their finest clothes, but they were suffering for it. They wondered why he was not dripping like them. It would have spoiled the illusion if he had confessed that just before their arrival he had changed into a new suit that one of his mechanics had freshly pressed in the balloon shed. As Santos-Dumont climbed aboard, double-checking each of the control wires, looking exposed with such a shallow basket around him, even Deutsch momentarily forgot the Brazilian's youthful impertinence and genuinely wished him well. The motor did not sound particularly good—it was coughing in the heat—but there was no backing out now that he had assembled the Aéro Club. Santos-Dumont was in the air by 6:41 A.M., and eleven minutes later he was within fifty yards of the tower. The judges were convinced that he would win the prize. But Santos-Dumont had his doubts—he knew a tailwind had helped him on the first leg and, unless the weather changed, that same wind on the return trip would slow him down. He rounded the tower at an altitude of 360 feet, the balloon glistening in the early morning sun. Sure enough, he was fighting a strong headwind, which interfered with his efforts to descend at the Parc d'Aérostation.

Ironically, it was Deutsch himself, or rather his constructions, that interfered with the landing. "The entrance to the park was obstructed by two very high sheds in which M. Deutsch, donor of the prize, is building a large balloon for the purpose of winning it himself," the *New York Sun* reported. "M. Santos-Dumont made repeated attempts against the wind to enter the park between the sheds. . . . The struggle lasted five minutes. The supply of petroleum became exhausted and the balloon was left at the mercy of the wind." In the interest of making a quick descent now that the engine had stopped, Santos-Dumont ripped the silk envelope, but before *No. 5* could deflate, it was blown backward a quarter mile across the Seine and he found himself entangled high in a chestnut tree in the garden of Edmund de Rothschild. This was the second Rothschild—he was Alphonse's younger brother—to whom he had paid an unexpected visit, and the hospitality extended to him was as great as the first time. Santos-Dumont knew that if he was going to crash-land, it was best done in the wealthiest part of town. Edmund de Rothschild's gardener put a ladder against the tree and climbed up to see whether Santos-Dumont was all right. "I'm thirsty," he told him, and soon servants rushed from the house carrying an ice bucket of champagne to wash away the trauma of the crash. The men also offered to extricate Santos-Dumont and his airship from the ensnaring branches. But the Brazilian asked them to hold off until he could devise a plan to free the balloon without damaging it further. He wanted to be alone with his champagne at the top of the tree.

As fortune had it, Rothschild's neighbor was Princess Isabel, Comtesse d'Eu, the daughter of Dom Pedro II, the last emperor of Brazil. When she learned of her countryman's predicament, she ordered her servants to prepare a sumptuous lunch and carry it up the tree in a picnic basket. They also delivered an invitation to join her for drinks.

After apologizing to Rothschild for knocking the nuts off his tree, he took the princess up on her offer. He had been wearing a bright red tie, and feared that the color might bring her sad memories of the revolutionary sentiment that had toppled her father, so he swapped the tie for a bystander's black one. "But for the turn of destiny which deposed Dom Pedro," the next day's papers chortled, "M. Santos-Dumont would have been a subject of the titled lady whose hospitality he so unexpectedly received." The countess's parting words evidently meant a lot to him because it was the only part of their conversation that he preserved in his diary: "Your evolutions in the air make me think of the flight of our great birds of Brazil. I hope that you will do as well with your propeller as they do with their wings, and that you will succeed for the glory of our common country!"

To the local press, it did not matter that Santos-Dumont had ended up in a tree. "Paris to-day," one correspondent gushed, "witnessed the official birth of an invention which may revolutionize the commerce of the world within the next few years." New York papers were equally effusive. "The hero of the hour is certainly M. Santos Dumont," the *Herald* announced. "Like Byron, he awoke one morning to find himself famous. His successful experiments in aerial navigation on Friday and Saturday have carried his name to the remotest corners of the world." But the occasion was also the first time that the press coverage was not completely favorable. A few American papers questioned the lighter-than-air approach. Under the headline "Balloon Navigation Impracticable," the *Chester Democrat* opined that

> a balloon will necessarily always be at the mercy of the air currents or winds when they exceed the velocity and power of mere zephyrs. If an aerial machine which can fly in the teeth of a gale is to be constructed it must be along the general lines followed by Professor Langley, whose self-propelled experimental machines secured levitation by means of their motors and were not dependent upon a huge gas-bag for their lifting power. There is more "revolution" embodied in one of Professor Langley's aeroplanes than in all the balloon airships ever constructed.

In the past, Santos-Dumont had responded to critics with bravado and exaggerated accounts of his flights. But this time he was modest and self-deprecating, and he stretched the number of years he had been in aeronautics from four to fifteen in order to underscore how little he had done: "The only thing I have accomplished in fifteen years of experimenting, during which I have wrecked four aeronefs, is to be able with tolerable certainty, in fine weather and with a mild breeze, to start from a given point and navigate through the air in any direction, right or left, up or down. To anything more than this I have no pretensions." That his confidence could be undermined so easily, and that his only responses were the two extremes of arrogant denial or groveling acceptance, was not a good sign that he could handle the nastier backbiting that would invariably come as he attained more success.

On Sunday, July 14, Parisians held their annual Bastille Day fete of fireworks, music, and marathon dancing. Wreaths were placed on the Strasburg Monument, in the place de la Concorde, by various political groups, although the police removed those left by the socialists, which were inscribed, "To the Fatherland's Victims." Open-air balls were held in almost every square, which were elaborately decorated with bunting and Chinese lanterns. Even in the poorest neighborhoods, "a few planks placed on barrels, or a wagon draped with the national colors and ornamented with half a dozen lanterns, served as a stand for a scratch orchestra, around which workingmen and their families dance the night long." All of Paris was a street festival, and only the most extreme behavior stood out. A girl who worked in a Montmartre music hall bet her friends that she would pass the night in jail. "She dined in the Bois and supped in Montmartre, and was therefore well-stocked with enthusiasm," the society columns reported. "She began by smashing glassware in a café. The police, seeing the woman with jewels, at first hesitated." She continued to break things but still failed to get arrested. To win the bet, she had to resort to punching a policeman.

Cars and bicycles were part of the Bastille Day celebration, but the airship, the transportation hero of the weekend, was conspicuously missing. Santos-Dumont had taken the evening off to be with his friends on the ground. He was not one for dancing but eating and drinking and taking in the revelry from the terraces of the cafés suited him just fine. At midnight, a parade of torch-bearing cyclists and automobilists made their way from the Bois de Boulogne to the Latin Quarter. At

Longchamp, President Emile Loubet reviewed a company of military cyclists who had just arrived in Paris after a three-day ride from Sedan, near the Belgian border. Like choreographed dancers, the cyclists demonstrated in unison what was special about their bicycles—they could be folded up in thirty-five seconds and then carried away on the backs of the men. At the ceremony, those who favored wheeled locomotion over the old-fashioned horse were cheered by the release of country-wide accident statistics for the month of March: the horse had caused seventy-seven deaths, the railway nine, the bicycle three, and the automobile three.

Santos-Dumont felt upstaged by all the Bastille Day activity that did not center on him. He started boasting oddly to those within earshot that he had fulfilled the terms of the Santos-Dumont prize the day before, even though, clearly, he had not flown "without touching the ground" and "by his self-contained means alone." This conceit was more than Henry Deutsch could bear; he wanted Santos-Dumont evicted from the Aéro Club. Santos-Dumont's friends, though, were more worried than exasperated.

Madness was the talk of Paris that July. It was front-page news when Dr. Gilles de la Tourette, the chief physician at the Paris Exposition and one of the country's leading experts on "mind troubles," was sent to an insane asylum. The French medical community, while entertaining hope for his cure, debated the cause of his "mental imbalance." Some of his colleagues believed that overwork had unhinged him. Others pointed to an injury he had sustained when a female patient, whom he had sent to an asylum, shot at him, the bullet just grazing his head and "preying on his mind." If a man as accomplished as Dr. Tourette—immortalized for his discovery of the neurological disorder known as Tourette's syndrome—

could suddenly go mad, then it seemed that no capable person was beyond insanity's reach. Santos-Dumont's friends knew that the Brazilian was a great storyteller, and they could forgive small fibs that added color to a tale, but this was a case of his making a serious claim that everyone knew was patently false. What troubled them was that he had apparently convinced himself that he had flown around the Eiffel Tower without touching down. The Santos-Dumont they knew would have been mortified to learn that people thought he was a liar. For now, though, his friends did not confront him, hoping that his brush with madness would go away like a cold, and indeed, it seemed to pass.

Santos-Dumont went to great lengths to craft his public image. He would have hated that his name and Tourette's were being mentioned in the same conversation. He subscribed to three clipping services so that he could monitor what the world's newspapers were saying about him. So far he had won reporters over not just by his courage and inventiveness but by his flashy wardrobe, his exotic tales of life in Brazil, the fancy dinners he treated them to at Maxim's, and his facility in their native tongues (he spoke French, Portuguese, Spanish, and English). His passion for ballooning was contagious, and reporters unwittingly conspired to depict him as a modern-day Icarus, as a romantic figure bent on conquering the skies.

The Bastille Day celebrations spilled into the following week, closing many businesses. When the festivities died down, Santos-Dumont began repairing the damage the chestnut tree had done to his airship. On the weekend of July 20, 1901, word spread through Paris that he was about to try again for the prize. The heat wave had continued, and fierce thunder and hailstorms hit the country and the rest of Europe. Temperatures in St. Petersburg reached 117 degrees, and a record number of people were killed by lightning in Germany, Austria, Holland, and France. Four children tried to hide from a storm in a church steeple outside Paris, but a lightning bolt killed them as they rang the church bells. The inclement weather did not stop hundreds of Parisians from camping out at Santos-Dumont's balloon shed. Word of his imminent take-off turned out to be incorrect, though, and even if the rumor had been true, the lightning and hail would have forced a postponement.

By the end of the month, the airship was fully repaired. Before trying again for the prize, Santos-Dumont ascended daily for quick jaunts. One trip went smoothly until the end. According to an eyewitness:

> The flying machine was almost above its shed in the Parc d'Aérostation, and the spectators, who had been watching its graceful evolutions and admiring the navigator's control of his huge craft, were waiting for the descent. Suddenly, Santos-Dumont was seen to clamber out of his basket along the slender framework supporting the motor. If he had slipped, if a sudden gust of wind had struck the balloon and caused him to lose his hold, he would have plunged downward three hundred feet to destruction. The spectators gasped and shuddered, and when the aeronaut regained to the basket in safety they cheered. One of the coupling wires had become jammed against the side of a pulley. It was a most dangerous thing to try to free it, but Santos-Dumont did not hesitate for a second.

Another time, on July 29 at 4:35 P.M., the motor misbehaved as it so often did, and he aborted the flight early, the guide rope cutting his fingers as he descended. When he landed, the crowd, "including a large number of women, whose handsome toilettes gave color to the scene," fussed over his bloodied hand. Only one spectator was ungracious, insisting that he should go right back up and fly again even if the motor was defective. Santos-Dumont climbed down from the airship and pointed to the seat, "Here is my place," he said. "Now you can try." The troublemaker, "to the amusement of the onlookers," the *Herald* reported, "retreated into the background." A few days later Santos-Dumont had to abandon yet another flight when the guide rope snagged in a tree.

Even as he was stumbling in his efforts to win the prize, French aeronauts, fearing that he might usurp their place in history, started a smear campaign against him. Chief among them was Colonel Charles Renard, who told the press, "M. Santos-Dumont is nothing better than a society sportsman of no scientific attainment." In 1884, Renard and Arthur Krebs, both members of the French army, had constructed *La France,* a sixty-six-thousand-cubic-foot balloon powered by an electric motor. On the first ascension, the balloon successfully returned to its starting point, the military balloon station at Chalais-Meudon. In twenty-three minutes *La France* had flown just shy of five miles. Renard and Krebs made six other flights, two of them over Paris, and only twice were they unable to return to Chalais-Meudon. The army was not enthusiastic about *La France,* however, because its heavy motor was so weak that it could fly only in calm skies. Indeed, from the moment it was completed and ready to fly, Renard and Krebs had to wait two months for a day with light-enough winds. Renard estimated that in still air its motor could propel it 14.5 miles an hour (Santos-Dumont could reach twenty miles an hour). Although Colonel Renard gave up on *La*

France after the seven flights, he was still in charge of aeronautical research in the French military at the start of the
twentieth century.

Colonel Renard regarded the Deutsch prize as a silly award
for a feat he believed he had achieved a decade and a half
before. In all likelihood *La France* would have been defeated
by the inevitable headwind that came after rounding the Eiffel
Tower, but Santos-Dumont's airship was not that much faster
and some of its most important features—its cylindrical shape,
its interior balloonet—could be found in *La France.*

Apart from Renard's contemptuous description of Santos-
Dumont to the press, he preferred to let surrogates besmear
the Brazilian's reputation. At the end of July, Renard's supporters pulled off a brilliant coup at the Aéro Club, most of
whose members had dispersed for vacation, by orchestrating
the nomination of Renard and his brother for election to the
Aéro Club's aerostation committee. As members of the jury
for the Deutsch prize, the brothers would be in a position to
frustrate Santos-Dumont's efforts. The Brazilian's friends in the
Aéro Club rallied to his cause. Wilfrid de Fonvielle, the chair-
man of the committee, objected to the nomination of two people so hostile to the aim of the competition and its foremost
competitor. In press interviews, Emmanuel Aimé, the club's
secretary, dismissed the importance of Renard's 1884 flights.
When Aimé returned to the club's headquarters, he found himself locked out of his office and stripped of the stipend that
came with the position of secretary. Aimé threatened to air the
club's dirty laundry if the padlock was not removed. As if to
offset Aimé's shabby treatment, the Aéro Club announced that
by unanimous vote it was conferring a gold medal on Santos-
Dumont, but he did not show up to receive it. Because the

minutes of the club's meetings were sealed, it was not clear who was siding with whom. The press relished the mysterious dispute, and ran headlines describing it as "another Dreyfus affair."

Santos-Dumont tried to stay above the fray. He retreated to his apartment for a few days and assessed what had gone wrong in his recent flights. He was cheered by news that the Brazilian government was considering funding his experiments. "It is a very kindly idea," he said, "and if it were carried out it would give me genuine satisfaction, not on account of the money, but because it would be so encouraging to feel that I had the sympathy and tangible support . . . of my countrymen. I think, also, it would be a good investment for Brazil, as it would attract public attention in Europe to the country, and in a very favorable way . . . for up to the present most people associate South America rather with revolutions than curiosity about scientific matters." The Brazilian government soon came through with a grant of fifty thousand dollars.

He also received a letter from Princess Isabel, along with a present she had commissioned from Cartier:

August 1, 1901

Monsieur Santos-Dumont:

Here is a medal of St. Benedict that protects against accidents.

Accept it and wear it, at your watch chain, in your card case, or at your neck.

I send it to you, thinking of your good mother, and praying God to help you always and to make you work for the glory of our country.

ISABEL, Comtesse d'Eu

Never one to follow the advice of others, even a princess's, Santos-Dumont dismissed Isabel's three suggestions for wearing the medal and instead turned it into a bracelet by adding a thin gold chain. The Saint Benedict jewelry became as much a part of his standard wardrobe as his panama hat, upturned shirt collars, and dark suit.

On August 8, at 6:00 A.M, wearing the Saint Benedict bracelet for the first time, and with the renewed support of the foreign press corps, Santos-Dumont made his next assault on the Deutsch prize. "We saw the balloon climb high into the heavens, and turned its yellow prow toward the Tower," the Paris correspondent for London's *Daily Express* reported. "Santos-Dumont came rushing through flight as straight as a rifle-barrel, and seeming as fast as a rifle-bullet." Never mind that he was traveling no faster than a good athlete could sprint, the correspondent was just warming up.

> His ship was sailing with the wind, and sailing faster than ever moved a Yankee cup defender. The rattle of the motor, heard more than a mile and half away, brought people to the roofs. The workmen gazed aloft. It was an exhilarating spectacle, thrilling, enchanting. Soon he was at the Tower. He rounded the great iron stake with ease, gracefully slowing down, so as to convince all of his performance; and as the ship swung around there was a mighty cheering from far and near. Santos-Dumont waved his hat in pleased response.

He had made it to the tower in a record nine minutes and rounded it in another thirty-four seconds. Again the judges thought the prize was within his grasp, but again Santos-Dumont recognized that all was not well.

Even before reaching the Eiffel Tower, he had suspected that the balloon was losing hydrogen through one of its two automatic gas valves, whose spring had been accidentally weakened. Ordinarily he would have descended at once to examine the valve. "But here I was competing for a prize of great honor," he recalled, "and my speed had been good. Therefore I risked going on." As he headed back toward Saint-Cloud, his suspicions were confirmed when the balloon started to sag. When he passed over avenue Henri Martin, a strong gust of wind hit the limp gasbag, throwing it violently back fifty yards. The suspension wires, which had slackened, were dangling ominously near the propeller. Santos-Dumont saw the blades cutting and tearing at the wires, and he stopped the engine at once. With no motor to fight the strong headwind, the airship was pushed back toward the tower. At the same time it shot upward until it was twice the tower's height. The *Express* correspondent was watching Santos-Dumont through binoculars: "Two thousand feet in the air he climbed from his basket and clambered along the dipping, swaying, gossamer-like keel. . . . It was sublime daring," as he worked to free the piano wires from the propeller. "Men below turned away their eyes. Santos-Dumont stood well outside the basket on two slender sticks, not so large as broom handles and three feet apart. He held on by the third strand of wood that formed the apex of the triangular keel. He worked furiously for a few seconds, and then went back to his car." The gasbag, now a quarter deflated, was rocking wildly. Like a ship in a storm, *No. 5* pitched and rolled.

When the bow rose up the gas swelled it full, and the stern crumbled, and doubled over empty. As the bow dipped this action was reversed, the stern filled and was buoyant, while the other end sagged like a wet towel. The ship fluttered in the air and flopped alarmingly. Santos-Dumont was being tossed on great wind swells in a wrecked airship! So that he would not fall out, he hooked a wire through his belt and secured himself to the ship. For a moment the balloon was so folded over that the propeller, which was still spinning, flicked the gasbag. There was a ripping and a rush of gas. With a sickening, despairing flop, the airship began falling.

"It may have seemed a terrific fall to those who watched it from the ground," Santos-Dumont said later, "but to me the worst detail was the airship's lack of equilibrium. The half-empty balloon, fluttering its empty end as an elephant waves his trunk, caused the airship's stem to point upward with an alarming angle. What I most feared, therefore, was that unequal strain on the suspension wires would break them one by one and so precipitate me to the ground. Why was the balloon fluttering an empty end and causing all this extra danger?" Why had the fan, before he cut the engine, not filled the interior balloonet and swelled out the gas balloon around it? At the time the only explanation he could think of was that the engine must have slowed, reducing the fan's power. That was strange, he thought, because he could generally tell by the sound of the engine if its speed had changed, and this time he heard nothing unusual. Later he learned from his workmen that the varnish sealing the interior balloonet may still have been wet, in which case the silk may have stuck together and not fully inflated. Santos-Dumont himself was at fault. He had been in too much of a hurry to wait for the varnish to dry.

The airship was falling faster, and at the same time it was still being blown backward. He could have jettisoned ballast to arrest his fall, but he feared that if he did not soon reach the ground he would suffer a worse fate by being "dashed against the tower." The Seine would be a forgiving landing place, if he could manage to get to it, a half mile away.

"One could hear screams from every housetop," the *Express* reported. "But Santos appeared in no degree alarmed. He slipped the heavy guide-rope down by the stern in order to lower that end—as he was in the bow it was better for him to have the stern hit the ground first. . . . The wind was blowing the ship so that it slanted down a long incline, whose end appeared to be the Seine. The last thousand feet were covered with a rush like that of a lift falling down its well."

He hoped to land in the river just beyond the Trocadéro Hotel, lodging that had been erected to accommodate tourists during the recent exposition. A rescue boat moved into position. To gain time to drift past the hotel, he started feverishly throwing out sand ballast, which was not easy to do given the speed with which he was falling. "The whole of the keel had already passed the Trocadéro," he recalled, "and, had my balloon been a spherical one, it too would have cleared the buildings. But now, at the last critical moment, the end of the long balloon that was still full of gas came slapping down on the roof just before clearing it! It exploded with a great noise— exactly like a paper bag struck after being blown up. This was the 'terrific explosion' described in the newspapers of the day."

Santos-Dumont found himself lying prone across the keel, his head and shoulders bobbing in midair, some forty feet above the ground against the eighty-foot-high wall of the hotel. The keel dropped jerkily down the wall a few feet until it was wedged at a forty-five-degree angle against the roof of a one-story restaurant below, No. 12, quai de Passy. "The keel, in spite of my weight, that of the motor and machinery, and the shock it had received in falling, rested wonderfully," he said. "The thin pine scantlings and piano wires of Nice had saved my life!" A man on the roof of the hotel threw him a cigarette, and Santos-Dumont, who ordinarily did not smoke because he regarded it as a sign of moral weakness, eagerly puffed away. The fire department finally arrived, lowered a rope from the roof, and pulled him to safety. Then the firemen proceeded to rescue the airship. "The operation was painful," he said. "The remains of the balloon envelope and the suspension wires hung lamentably, and it was impossible to disengage them except in strips and fragments!"

Thousands of people who heard the explosion gathered at the hotel, where the police held them back with hastily erected barricades. "The reception of M. Santos-Dumont, when he reached the street was most enthusiastic," the *Herald* reported, "many women clinging to his neck and kissing him repeatedly." He displayed his medal of Saint Benedict to the crowd, kissing it reverently, and attributing to it "his narrow escape from death." Everyone could see that he was not even scratched, "and, as he superintended the removal of the machine, he said he was quite ready to begin operations again."

Deutsch, who had rushed to the Trocadéro, was ready to forgive all. He had actually shed tears when he saw the balloon falling. "He was so affected by the danger that M. Santos-Dumont passed through," the *Herald* said, "that he told him he would rather present the prize to him at once than see him kill himself with his experiments, but the balloonist replied that he had rounded the Eiffel Tower in such a short period of time that he considered the result too satisfactory to permit the relinquishment of future attempts." As if to underscore his determination, Santos-Dumont requested some petroleum. Then he revved up the two-hundred-pound motor, which the firemen had just hauled down from the roof. "M. Santos-Dumont looked and listened with evident pleasure as the flames escaped from the pipes," London's *Daily Telegraph* reported. "And the deafening explosions resounded with a healthy vigour, showing that the engine was undamaged. I think the picture of M. Santos-Dumont trying over his motor this morning half an hour after his fearful smash-up shows what kind of man the plucky Brazilian is." ("Plucky" was the adjective that the press universally used to describe him.) The crowd cheered Santos-Dumont's resiliency. Deutsch observed that the tattered balloon silk was scattered over the roofs of the neighborhood. A few weeks earlier, he had ordered his own cigar-shaped balloon—it would be huge at seventy thousand cubic feet. He was set to receive it any day, and he magnanimously offered it to Santos-Dumont. The Brazilian politely declined and gathered up whatever pieces of silk were in reach. Despite his praise of Santos-Dumont's courage and ingenuity, Deutsch did not actually believe that he would ever win the prize. "I am afraid the experiments will not be conclusive," Deutsch told the assembled reporters after the Brazilian had departed. "M. Santos-Dumont's balloon will always be at the mercy of the wind, which is not the kind of airship we have dreamed of."

•

A MAN of less resolve who fell from the sky might go straight
to bed or take up the bottle. Santos-Dumont went directly from
the Trocadéro to his workshop, where he examined what he
had managed to salvage from *No. 5*. The framework, which
had surprisingly withstood the crash, was damaged only when
the firemen extricated it from the hotel. The silk envelope fared
worse. He tested the largest piece of cloth he had recovered
with a dynamometer of his own invention to see how much
pressure it could withstand. The test showed that after what
the silk had gone through it was too weak to be incorporated
into another balloon envelope. Within hours of the accident,
he visited the balloon makers and placed an order for a new
airship, *No. 6*.

That night he dined at Maxim's, where he regaled the other
patrons with details of his crash at the Trocadéro. Most of the
diners would have listened for hours, but one woman complained that the discussion was "rather technical." And so the
conversation drifted to other topics of the day: Like the price
of absinthe, which had just gone up 30 percent because of the
destruction of a large factory in Pontarlier where a single brand

of "the little green goddess" was manufactured. Like the new spittoons, which looked like jam pots, hastily installed throughout the city, with the inscription *"crachior public,"* after the chief of police had banned spitting in the open air for fear that saliva spread tuberculosis. Like the infighting between two groups of animal lovers, the Société Protectrice des Animaux, which was distributing free hats to horses to protect them from the sun, and the Société d'Assistance aux Animaux, which declared that equine hats were noxious and that what horses really needed were parasols. Santos-Dumont was not the only hero of the day. A laborer named Simon was repairing a well near Chartres, when the walls collapsed and buried him. A team of engineers worked 117 hours before reaching him. They found him weak but conscious, standing upright with one arm raised, his back to the well shaft. Simon curiously insisted that he had been entombed for only twenty-four hours.

Santos-Dumont's fame was garnering him invitations from the United States. Organizers of the Pan American Exposition in Buffalo, New York, a rival to the Paris Exposition, offered ten thousand dollars if he would fly his airship around their Electric Tower for a distance equivalent to that required to win the Deutsch prize. The *New York Journal* offered to fund him to circumvent the globe in a lunar month or to fly to the North Pole. "I confess," he responded, "that the idea of breaking all records by making a tour of the world in less than a month, in other words, racing the moon, has an originality about it which captures the imagination, yet, as a student of science, I should prefer a trip to the Pole." Nothing ever came of these offers, though. Nor did he take aloft any of the numerous strangers who had invited themselves on his flights. "It appears that one of the greatest troubles of M. Santos-Dumont and other aeronauts," reported the *Herald*, "is to resist the pleadings of Paris actresses desirous of accompanying them on their ascents. This they regard as a short road to fame. Some of them even drop into poetry. M. Santos-Dumont probably owes his salvation to the fact that his balloon can only transport one person."

At the end of August 1901, Santos-Dumont received a summons to appear in the Court of the VIIIth Arrondisement to answer a suit seeking 155 francs (thirty dollars) for the damage to roof tiles on a building next to the Trocadéro. The building's owner, the widow Mme. Deniau, did not claim that Santos-Dumont's airship had actually smashed the tiles. Rather she maintained that they were broken by zealous bystanders who swarmed her roof in their eagerness to help the stranded aeronaut. But she was holding him responsible. The newspapers mocked the woman for having the audacity to sue the beloved inventor over such a petty matter. Santos-Dumont did not contest the suit. The court handed him the bill:

Subpoena	.50 franc
Summons	4.80 francs
Damages	150.00 francs
Stamp	.10 franc
Total	155.40 francs

He paid on the spot.

Of course he could easily afford the 155 francs—for him the issue was the uncertain cost of future mishaps. Did he need to build into his expenses "the vexations of who knows what prosecutions for knocking down the chimney pots of a great capital on the heads of a population of pedestrians?" He tried all the insurance companies in Paris but none offered to protect him against the damage he might do to others "on a squally day." Nor would they insure the airship itself against destruction.

He nevertheless proceeded with the construction of *No. 6*. At 108 feet, it was ten feet shorter than *No. 5*. But it was stockier—a fatter cigar—with a gas capacity of 22,239 cubic feet, a fifth more than *No. 5*. The propeller was at the stern again, where it would push *No. 6* through the air. As he planned a renewed assault on the prize, Deutsch thwarted him by suddenly changing the rules. Santos-Dumont learned of the changes secondhand and responded by making his case in the press, hoping that an outpouring of popular opinion would force Deutsch to revert to the original formulation. On September 11, the papers printed Santos-Dumont's indignant reply to the club's aerostation committee:

Monsieur le President:

I have learned through the press of the decisions come to by the Aerostation Committee of the Aéro Club at its sitting on September 7, namely, that the committee had decided that heretofore every competitor must not only return to his starting point within thirty minutes after having doubled the Eiffel Tower, but must also touch the ground within the enclosure of the park of the Aéro Club; that the time of trial will begin from the moment when the guide rope or any cord holding the balloon is loosened, and that it will end when the rope is seized by a man stationed in the enclosure.

Allow me, Mr. President, to express my astonishment at these changes. I refuse to believe that in the midst of a period of competition the Aerostation Committee of the Aéro Club wishes to add to the great difficulty of the trial, which already contains many, as I have proved at risk to my life.

The original text of the rules in force up to this day sets forth that the aeronaut must return and not that he "must come to the ground at the starting point." This liberal text, inspired by a sincere desire to see aeronauts carry out the programme, was adopted by the committee to soften the conditions imposed on making a trip from the Parc d'Aerostation to the Eiffel Tower and back in thirty minutes, a condition complicated by the obligation of announcing twenty-four hours in advance the attempt, thus making it subject to the uncertainty of the weather.

At the period when this text was settled the grounding of a steerable balloon returning at full speed from the tower to the park was just possible, in spite of the difficult crossing at the valley of the Seine, in which a current of moist air disturbs in a moment the aerostatic equilibrium. By skillful steering one might have attempted to approach on the side of the aqueduct carrying the water of the Avre, in spite of the narrowness of the ground between the trees and the houses on the bank, between telegraph wires and cables conveying electric currents, tightly stretched, between the tramway line and the railway line.

In the opinion of all aeronauts whom I have questioned this resource has just been prevented by the construction of M. Deutsch's balloon shed, twenty-seven mètres high and sixty mètres long which forbids access to the park to a steerable balloon coming at full speed on the only side which was at all practicable under the atmospheric conditions laid down.

If the aeronaut flies at full speed it is impossible to descend on the broken ground of the park. If he advances slowly he runs the risk of being carried by the wind along the banks of the Seine. I know this from experience, for I have already twice got broken thus.

And it is at this moment, on the morrow of my last accident, that the Aerostation Committee of the Aéro Club wishes to impose upon me the additional obligation of descending on ground where impossibilities are added to impossibilities, since navvies have been digging large trenches there. My workmen have already run risks in these trenches during the manoeuvres at the start. To ask them to seize my guide rope when I was passing over what was a park would be to expose them to dangers in which I will know their devotion would inevitably involve them.

Moreover, on race courses the timekeeper marks the time at the moment when the jockey passes the post, and not when, having stopped his mount, he hands the reins to the stable boys. Why should aeronauts managing a balloon, which from its mass when in movement represents an enormous living force, be compelled to have their guide rope seized as it passes and be stopped suddenly at the goal post?

The original text of the regulations is the only admissible one in the actual state of aeronautic science. I accepted it, and I kept to it, leaving to rasher people the business of decreeing more difficult conditions and that of carrying them out. If, therefore, I fulfill the conditions of the Grand Prix within the time specified, I shall content myself with simply passing over the park to establish my return in accordance with the rule to which I gave my adhesion, and should my guide rope touch the ground at that moment I forbid any workmen in advance to stop me as I pass, reserving to myself the time for returning and grounding wherever I please.

In 1899, before the foundation of the Grand Prix, I made evolutions round the Eiffel Tower in my third steer-

able balloon. I have been continuing since without concerning myself with the extremely arbitrary conditions of the competition, and shall continue to pursue methodically my series of experiments, which will only end with my life.

Nevertheless, as hitherto, I shall do my utmost to obtain the official approbation of the Aéro Club Commission.

I hope there will be a few unbiased witnesses around to attest the fact.

In default of official control, I shall be satisfied with that of the press, whose aid is of such value to the progress of the idea of aerial flight. After that, should the Grand Prix not be awarded to me, assuming that I have fulfilled the conditions, I shall regret it all the more because it has never entered into my head to receive the money. Just as I abandoned the interest—4,000f.—last year to the Aéro Club for the foundation of a new prize, I have devoted in advance the sum of 100,000f. half to the poor of Paris and the other half to the disinterested men who have been displaying devotion for which at times they have had to suffer.

I hope, in the interest of the poor of Paris and the men who have aided me, the Aerostation Committee will reverse its decision and allow me some sort of chance of winning the 100,000f. for them.

In any case my balloon will be in proper condition again by the end of the week and I hope to be able to continue my trials on Sunday next.

The Aéro Club ignored the letter, and Santos-Dumont redoubled his efforts to improve *No. 6*. The balloon had been out of commission since September 6, when the airship scraped a house after the guide rope became tangled in telegraph wires. Although the roof tiles were unscathed, the airship was damaged, its seventy-square-foot rudder broken and the silk ripped. On the morning of September 19, Santos-Dumont went up again in the newly repaired *No. 6*. The skies were calm although disconcertingly foggy. If he tried to ascend above the dense fog, he feared that as soon as he emerged, the sudden exposure to the sun's unshielded heat would quickly warm and expand the hydrogen, skyrocketing the balloon much higher. To regain control of the airship, he would be forced to vent hydrogen, which he did not want to do at such an early stage in the flight; later he might regret that he did not have enough gas. And so he chose to fly within the fog, 150 feet above the ground. The flight began auspiciously, but as soon as he was over Longchamp the ever-capricious motor started to sputter. While he waited for the motor to recover, he made tight circles over the racecourse, whose grassy center would make an excellent landing place in the event of an emergency. But he took one turn too sharply and clipped a stand of trees. Again the balloon ripped and disgorged the hydrogen. He was fortunate that he did not have far to fall. "The frame bearing the weight of the motor broke as soon as it touched the ground," the *Herald* reported, "but the aeronaut, doubtless protected by the medal of Saint Benedict . . . remained standing in the car, uninjured, in the midst of splinters of wood and strips of torn silk and a tangled web of steel and wire."

In all the previous accidents, the violence of the wind had played an important part. This time it had no role. It was the first mishap caused solely by pilot error. His supporters put the best face on it. "In aeronautics," Emmanuel Aimé told the press on his behalf, "especially is verified the old adage: 'Experience is better than science.' Inventors who content themselves with what is called 'room aerostation' have no idea of the difficulties of 'aerostation in the air.' M. Santos-Dumont has the great merit of seeking in his repeated efforts information of which his imitators—who it is to be hoped may be numerous—will have the benefit without having the trouble and expense."

Santos-Dumont himself was not bothered by what the papers delicately called "a mistake in steering in a moment of inattention on the part of the aeronaut." "Such accidents I have always taken philosophically," he said, "looking on them as a kind of insurance against more terrible ones. Were I to give a single word of caution to all dirigible balloonists, it would be, 'Keep close to earth!' The place of the air-ship is not in high altitudes, and it is better to catch in the tops of trees . . . than to risk the perils of the upper air without the slightest practical advantage!"

The Aéro Club still was not backing down on the eleventh-hour rule change, but public opinion was firmly on the aeronaut's side. News of the standoff had spread to far-flung regions of the globe. "M. Santos Dumont is to be pitied," the *Rangoon Gazette* observed:

After six attempts and the construction of three balloons to demonstrate the steerability of his airship, the Aeronautic Club intends to place spokes in his wheels. He is a Brazilian, he will not become a naturalized Frenchman, and he has never concealed his admiration for England. He is contesting the prize of 100,000 francs offered by a Frenchified German to whoever will succeed in steering a balloon from the Park of St. Cloud, round the Eiffel Tower, and back to the Park again. . . . Morally, Santos has accomplished that feat, but an ugly majority of the judges are jealous of the Brazilian winning all the glory from France. They have made a new condition, that the balloon must return to the "courtyard" of the Aeronautic Club, which is a very torn up piece of ground that might wreck his balloon.

Rule change or not, Santos-Dumont was prepared to make another run at the prize. He had fixed the motor after concluding that its erratic behavior stemmed from the fact that, being essentially an automobile engine, it was not designed to perform at all the inclinations an airship finds itself in. He changed the configuration of the carburetor so that no matter what its position the engine would never be short of petroleum. Moreover, the oil was now dispensed from four containers instead of one so that the motor would remain properly lubricated at any angle. He changed the location of the interior balloonet from one end of the outer balloon to the center, where it could more evenly and effectively prop up the tautness of the outer balloon. And the tiny valves that had failed so spectacularly two months before were replaced with the highest precision ones he could find.

On the afternoon of October 10, he took *No. 6,* with its reconstituted engine, for a spin over Longchamp. In spite of a contrary wind, he maneuvered the balloon above the racecourse for more than an hour in every direction "with perfect docility," under the watchful eyes of the Comtesse d'Eu and a host of other dignitaries. At 3:00 P.M. he brought the balloon down outside his favorite lunch spot in the Bois, La Grande Cascade, where he treated the countess and her husband to a quick drink. (The restaurant is still in business today, with the same Napoleon III furniture it had when it opened in 1865.) Fifteen minutes later, Santos-Dumont wound his way through the well-behaved crowd that had assembled to admire *No. 6,* which was parked like a very long carriage in front of the restaurant. He returned to Longchamp in the airship, and, crossing the Seine at a height of about 650 feet, slightly overshot the Parc d'Aérostation. With a sudden turn, he came back to his goal, and the airship, after brushing a telegraph wire at the western end of the park, hovered above the narrow space between his balloon shed and Deutsch's. Then he "wheeled around in circles," said Aimé, "like an eagle about to pounce on its prey. At one instant it was feared that he was lost, for he passed within two mètres of M. Deutsch's balloon-house, which blocks the entrance to the park on the south side. But he escaped the danger with a bold manoeuvre, that brought him exactly in front of his own construction, which he entered without waiting for the arrival of his workmen," who had been trailing him in an automobile at "express speed."

For the next week Santos-Dumont had given the Aéro Club notice that he would try for the prize every day, even though he had no intention of flying that frequently. The club disapproved of the blanket notice, but Santos-Dumont felt he had little choice. The rules of the prize required his notifying them twenty-four hours in advance, but he knew he could not accurately judge the weather a day before the trial. And so he summoned the jury daily in the hope that the skies would be favorable one of the times. At least he was no longer scheduling ascensions at the duelist's hour. After a week of aborted trials, the crowd at his balloon shed had largely dispersed. On Saturday, October 19, only five members of the twenty-five-member aerostation committee—Henry Deutsch, Comte Albert de Dion, Wilfrid de Fonvielle, Georges Besançon, and Emmanuel Aimé—showed up at the civilized hour of 2:00 P.M. Santos-Dumont called the Central Meterological Bureau at the Eiffel Tower and learned that the wind at the summit was blowing southwesterly at a speed of 13.5 miles an hour. He decided to ascend at once, in the presence of fewer than a dozen spectators. At 2:29 P.M. he took off, but in the haste in which the balloon had been prepared, it had been given too much ballast. As he left the park, the guide rope of the weighted-down airship caught a tree. He was forced to end the flight because he could only free the rope by descending. The sight of his airship, which had been visible to afternoon strollers across the Seine in the Bois, drew an enormous crowd to Longchamp.

At 2:42 P.M. he took off again, rose 250 yards, and headed straight toward the Eiffel Tower. The only glitch on the flight to the tower came when he passed over the Seine. "When the airship reached a point just above the river," he recalled, "it suddenly caught in a circular current of wind and lurched violently leeward. By quickly maneuvering the rudder and increasing the speed of the motor I was able to rectify the course almost immediately." The band of the Twenty-fourth Regiment had been marching up the Champs-Elysées serenading the visiting king of Greece and five hundred other dignitaries when someone shouted "Santos-Dumont!" and pointed to the sky. The band members dropped their instruments and joined, in the *Herald*'s words, "a wild stampede of foot-passengers, cabs, automobiles, and cyclists racing toward the Champ de Mars." Five thousand people reached the Trocadéro Gardens just as the airship, helped by a favorable eighteen-mile-an-hour wind, rounded the tower's lightning rod at a precariously close distance of forty feet. When the timekeeper at the tower announced that Santos-Dumont had set a mid-flight record of eight minutes and forty-five seconds, "many present literally danced with joy, and perfect strangers shook hands and congratulated each other as if it were a day of national rejoicing."

On the homeward journey, the twenty-mile-an-hour headwind rocked the airship and slowed it down, but *No. 6* was still making good time on a straight path to Saint-Cloud. Five hundred fifty yards from the tower, over the Bois de Boulogne, the engine stalled despite its new carburetor and lubrication system. At the risk of being swept off course, Santos-Dumont had no choice but to take his hands off the rudder controls, fiddle with the carburetor and ignition, and restart the engine. He lost a precious twenty seconds. The balloon passed over the Champ de Mars, close to the Hôtel des Invalides and Napoleon's tomb. "Here was another conqueror," gushed a London paper, "but a peaceful one this, a Napoleon of the air. It was no wonder that when some rays of autumnal sunshine appeared people cried out, 'The sun of Austerlitz!' " But before Santos-Dumont could win the prize, he had to tame the capricious engine. It stalled again as he crossed the fortifications that demarcated the city of Paris, and he easily restarted it.

A third engine failure was more serious. The airship fell rapidly as the propeller slowed. He managed to fix the engine while at the same time pitching considerable ballast to arrest the balloon's descent. With the balloon's equilibrium restored, he could concentrate on steering *No. 6* home. "The rest of the trip was remarkably intoxicating," he said later. "The four cylinders worked charmingly, everything was in shipshape, and I felt like putting my hands in my pockets and letting the thing travel on." As he crossed the Seine, he looked down at the bridges and crowds on the bank. He heard "wild cheers blended in one vast clamor. I said to myself that this must be a good sign, indicating that I was on time, but I had no watch and really could not tell anything about it. When the park came into sight, I shifted by sliding weight forward, causing a slanting plunge downward, because I did not want to arrive at a point too high. The airship obeyed the rudder so well that I was able to pass exactly in the centre of the grounds of the Aéro Club." As he overshot the starting point, the official timekeeper clocked twenty-nine minutes and fifteen seconds. Another minute and twenty-five seconds passed while Santos-Dumont turned the airship around and brought it back to the starting point, where his workmen grabbed the guide rope and reeled him in. When his basket was low enough for his voice to be heard over the applause, he leaned over the side and yelled, "Have I won the prize?"

Hundreds of spectators responded in unison, "Yes! Yes!" and swarmed the airship. He was showered with flower petals that swirled like confetti. Men and women cried. The Comtesse d'Eu dropped to her knees, raised her hands to the heavens, and thanked God for protecting her fellow countryman. The countess's companion, the wife of John D. Rockefeller, squealed like a schoolgirl. A stranger presented Santos-Dumont with a small white rabbit, and another handed him a steaming cup of Brazilian coffee.

Santos-Dumont was smiling broadly when the one dour face among the celebrants approached him. Comte Albert de Dion could not bring himself to look Santos-Dumont in the eye as he shook his hand. "My friend," he said, "you have lost the prize by forty seconds." Dion reminded him that under the amended rules the race was over not simply when he returned to the starting point but only after the further action of his men grabbing the guide rope.

"Nonsense!" the crowd protested.

Dion repeated his pronouncement that Santos-Dumont had not won. "That is the decision of the committee," he said, "in accordance with the rules of the contest."

Santos-Dumont offered to repeat the flight on the spot, but

the crowd would not let him. "You don't have to prove anything," they shouted. "You've won! You've won!"

He stood up in the basket and addressed the group. "As for
the prize," he said, "it is little to me not to receive it. It is the
poor who will suffer!" A few fists were shoved into the air. How
dare the wealthy car-owners who controlled the Aéro Club ca-
priciously deprive the city's indigent of what was due them!
Deutsch himself, who shifted alliances as easily as the wind,
stepped forward and momentarily defused the tension. He embraced Santos-Dumont and declared, "For my part, I consider
that you have won the prize!" While the multitudes cheered
anew, Dion managed to slip out of the Parc d'Aérostation in one
piece. Deutsch, unaccustomed to ovation and enjoying it, offered to give twenty-five thousand francs of his own money to
the poor if the committee refused to reverse its decision.
Santos-Dumont, though, turned down the offer. He was not
going to sell the poor short, he declared. Fists went up again,
more of them and much more firmly. The police moved in
and dispersed the crowd.

A reporter asked Santos-Dumont why he had passed over
the starting point. "I could have landed," he replied, "for I
have done it fifty times. I went onward because I expressly
wished to show the committee my independence of its whim-
sical and arbitrary decrees of a few weeks ago, when it was
decided that I must not only pass over the starting point but
land. So today I overshoot the mark as a race horse overshoots
the line at the track." It was his defiance more than his success
that particularly angered Dion and other senior members of
the Aéro Club. Always the chameleon, Deutsch now conceded
that while Santos-Dumont had clearly achieved a "moral victory," he "may not have materially fulfilled the conditions of
the prize." The Brazilian was done arguing, and the stress of

flying—the engine failing three times in mid-flight—had finally caught up with him. He climbed down from the basket, got into his car, and headed home. "When M. Santos-Dumont's well-known little electric automobile turned into the Champs-Elysées," the *Herald* reported, "it was escorted by hundreds of cycles and automobiles, and all the way down the avenue it was a triumphal progress. People on the sidewalk, in cabs, carriages and omnibus waved hats and handkerchiefs and cheered him . . . till he disappeared within the gate of his residence."

On Sunday, while Santos-Dumont waited to see if the prize was officially his, Gustave Eiffel invited him to lunch in his small private apartment at the summit of his tower. Although Santos-Dumont had circled the tower seven times in a balloon in the past four years, he had not set foot in it since he first visited Paris more than a decade ago. After lunch, Prince Roland Bonaparte, the president of the aerostation committee, sent him a congratulatory telegram. "So far as I am concerned," the prince said, "you have won the prize. I do not want to prejudice in any way the decision of the committee, but I consider that M. Deutsch should not pay you the 25,000f. which he offers you, but that he owes you 100,000f. I congratulate you warmly on the noble use you intend to make of the money." When the text of Bonaparte's telegram was published in the papers, which were devoting pages to the fracas, the waffling Deutsch announced that he hoped the committee would allow him to pay Santos-Dumont the hundred thousand francs. Deutsch could have it both ways because his vote, when the aerostation committee met in two days, would be cast in secret.

If it had seemed an exaggeration before to call the internal disputes at the Aéro Club "another Dreyfus affair," the description was now more apt. Whether the Aéro Club would back down was "the burning question in Paris," the *Herald* reported. "Political feeling even is beginning to manifest itself. M. Rochefort and M. Drumont made a violent anti-Semitic attack on M. Deutsch, making him responsible for withholding of the prize award, while partisans of the military balloon park, of Meudon, who have been against Santos-Dumont from the beginning, naturally still continue their opposition."

In the minds of ordinary Frenchmen, there was no doubt that he had won the Deutsch prize. "Santos, the great name of the week and the year, a name which the telegraph wires have already carried to the four corners of the universe, [is] the god of Paris, and his name rhyming with those of the popular heros, Pothos and Atmos [two of the Musketeers], will be heard to satiety," crowed one local paper. "The tailors, the pâtissiers, and the makers of new toys will raise it to immortality."

"The latest fashion from Paris in lady's hats was the 'Santos-Dumont veil,' " New York's *Dry Goods Economist* reported. "It was adorned with small velvet appliqués which had the shape of Santos-Dumont's dirigible balloons."

The sweet in demand on the streets of Paris was his gingerbread likeness. "Even the tiniest toddlers are heard to stammer, *'Un Santos, s'il vous plait,'* to the sellers of cakes under the trees," the *Herald* reported. "A story exists according to which a former President of France asked an official, 'Am I really popular?' to which he received the reply, 'not yet, sir, your figure in gingerbread is not being sold on the Champs Elysées.' "

The toy makers changed the designs of their balloons overnight. "It is a sign of the times that the toy balloons which are distributed by many Paris stores have a different shape," the

Herald said. "They used to be spherical but now they are cigar shaped and they have the name 'Santos-Dumont' written on them in bright colours." A tabletop dirigible was also a hit. The tiny airship could actually be flown by filling it with coal gas. But it was quickly discontinued because parents deemed it unsafe. The stores replaced it with a miniature version of *No. 6,* selling twenty thousand in three months. The toy was billed as a true flying machine, although the closest it came to flying was to be pulled through the air by a wire. "The fact that it cannot fly doesn't affect either its name or its popularity," the *Denver Times* reported, and then cheekily added: "Make-believe, therefore, must still play a part in the successes of the nursery flying machine, as it does very largely in that of the grown-up affair."

When the aerostation committee finally convened, on Tuesday, October 22, its decision pleased no one. The committee announced that it was postponing until November the question of whether Santos-Dumont had won the prize, but in the meantime it was reopening the competition. If another aeronaut flew around the Eiffel Tower before the end of the month, he would share in the hundred thousand francs or perhaps keep it all. In anticipation of a favorable decision, Santos-Dumont had asked the prefect of police to distribute the money on his behalf to the poor. Now thousands of beggars swarmed around the police station demanding their rightful share. Wealthy Parisians, fearing that class war might break out if the angry mob turned on their villas, came forward with large donations. Deutsch himself contributed twenty-five thousand francs (five thousand dollars), and a philanthropist named Daniel Osiris, whose family owned Empress Josephine's old chateau, offered to give Santos-Dumont a hundred thousand francs if the Aéro Club refused. A week later, on November 4, the committee succumbed to the public pressure and voted thirteen to nine to award Santos-Dumont the prize.

But the action was too late to appease him. He promptly resigned from the Aéro Club, thanked the people of Paris for their support, and announced that he would be spending the winter in Monte Carlo, a place, he said, where the "authorities" fully embraced aerostation. As for the prize money, *Le Vélo* reported, "the press had made Santos-Dumont's face so well known that when he went to the Credit Lyonnais Bank with a cheque payable to M. Santos-Dumont, the clerks did not hesitate in giving him 100 notes of 1,000 francs, without requiring any identification." In the end he gave twenty thousand francs to his loyal supporter Emmanuel Aimé, thirty thousand francs to his workmen, and fifty thousand francs to the poor. He asked the prefect of police to use the fifty thousand francs to redeem from pawnshops necessities like tools and furniture and return them to their owners.

"The Parisian populace must always have a hero, an idol of some sort, and to-night the young aeronaut occupies the pedestal," a British journalist wrote on the eve of the disposition of the prize money. "One knows not what to admire the more, the splendid courage of this daring youth from beyond the seas or the inventive genius that made such a feat possible. He was popular enough before, but his gift to his faithful assistants and the poor has raised Santos-Dumont to the pinnacle of adoration by the people of Paris."

Santos-Dumont's friends were not about to let him leave town without a grand send-off. It did not seem appropriate to have a banquet restricted to the usual Maxim's crowd of noblemen and playboys. After all, it was the people of Paris, not the noblesse, who never wavered in their support for him. And so his farewell dinner on November 9, 1901, at the Elysée Palace Hotel was open to anyone who paid twenty francs. His friends were a bit naive to think that anyone but the prosperous could afford that much for a meal. Nonetheless, the 120 people who showed up to dine on "aerial sole, very delicate and light as required, a Brazilian ice received with great applause, and a Santos-Dumont basket of fruit" were an eclectic group. "Princes and mechanical engineers, wealthy amateurs and retiring men of science, had all assembled," the *Daily Telegraph* reported, "with no other thought save admiration and friendship for the pluckiest and most modest of all the pioneers of aerial navigation." Most of the Aéro Club members stayed away but Deutsch attended. He had composed an after-dinner waltz called "Santos," which he instructed the Neapolitan band to play. The piece was a hit, and Santos-Dumont clapped enthusiastically. For an encore, the band performed a lively rendition of another of Deutsch's aeronautical compositions, the "Montgolfier March." Princess Isabel sent a giant chrysanthemum arrangement in the shape of *No. 6* and the colors of the Brazilian flag to the aeronaut's table. Gustave Eiffel presented him with a gold medal that showed him circling *his* tower. The painter Balaceano unveiled a large watercolor depicting Santos-Dumont hovering in *No. 6* and emptying ballast bags filled with thousand-franc notes. Cigars were lit throughout the banquet hall. After a few puffs, the celebrants started waving their cigars dangerously close to the table linens, in mock defiance of the warning Santos-Dumont had been given that sparks from his petroleum motor would explode his balloon.

Before departing for the Riviera, Santos-Dumont visited London. The Aero Club of the United Kingdom had just been founded by C. S. Rolls and other British automotive pioneers who were inspired by the Brazilian's successes. Santos-Dumont was the honorary founder of the club and a congratulatory dinner was held on November 25 in the Whitehall rooms of the Hotel Metropole. Knowing their guest's penchant for fine food, the hosts fretted over the menu and settled on nine courses built around his favorite dish, fillet of sole. Between the cheese fondue and the petits fours, Colonel Templer, head of military ballooning in the United Kingdom, toasted Santos-Dumont and admitted that he and his fellow army balloonists had all thought the wind was too rough for him to succeed in circling the Eiffel Tower. They were astonished, he said, when he not only made it around but did so within the prescribed time. "When M. Santos-Dumont stood up to respond," the *Daily Messenger* reported, "the whole of the guests acclaimed him with a warmth which belied the character of phlegmatic that has been given to the English. Table-napkins were waved, and for some minutes there was a veritable tempest of applause. Then the room burst into 'For he's a jolly good fellow.' " And that was before Santos-Dumont made a toast to the "Great British nation, which, after gaining the Empire of the Seas, aspires to the Empire of the Air."

The senior members of the British press corps had attended the dinner to interview the world's most famous aeronaut. They were divided on whether he looked like an intrepid flier. The reporter for the *Daily News* thought he did: "If one could create an aeronaut, he would make him like M. Santos-Dumont—rather below the middle height, lightly built but wiry and full of fire. If such a man fell out of a balloon one could easily imagine that he would be hurt less than others. But although he speaks our language fluently, an Englishman would not mistake him for a countryman. His jet-black hair, dark eyes, and swarthy skin do justice to his native land."

The *Brighton Standard* painted the opposite picture:

Santos-Dumont is the very last person in the world one would take to be the fearless and reckless man who has often faced death in the pursuit of scientific discovery. Rather below middle height, with a very slight and almost boyish figure: with a boyish face—long, narrow, with black hair worn down rather low on the forehead, he looks a little like a Brazilian version of Phil May. And his manner is quietude itself. As he discusses ballooning as one might the weather one finds it almost impossible to realise that he is in the presence of a man whose name and fame have been made immortal; but only by exposure of his life over and over again to death in a horrible way.

The newspapermen took advantage of the dinner to question the aeronaut about his experiments. The one journalist in the room who did not know much about his previous flights asked him if he had ever met with any accidents besides the one at the Trocadéro. "Yes," he laughed. "I have had many accidents, useful ones. They all taught me something, but I was never hurt except once. It was at Nice. The skin was scraped off my face, but you cannot see any mark now. I was—what do you call *trainé*? Dragged? Yes, I was dragged along on the ground. As soon as the balloon started a hurricane broke." That much was true but he embellished the danger by adding the fictitious detail that torpedo boats were dispatched to shoot the balloon down but held their fire because he did not drift out to sea. "I went along the land," he continued, "and was dragged till the balloon burst against a tree."

He was also asked whether he had any objective in coming to London besides the dinner in his honor. "Yes," he said. "I also come to London to see whether I could have a few balloon trials here. . . . I feel it would be perhaps less dangerous for me than Paris, the houses not being so high. Yet there is a danger here. I see many wires over the town. There are none in Paris, and wires may cut a balloon. Yet I hope next year I shall have a few trials in England. I do not know. Your climate? Ah! Well, I know it." Everyone chuckled, and before the evening was over rumors were circulating that he planned to fly around St. Paul's Cathedral. He did nothing to discourage the speculation, and when he headed back to France a few days later, the English were convinced that in the spring Santos-Dumont would move his ballooning facilities to London.

[C H A P T E R 8]

"MAKING ARMIES A JEST"

•

A S A consequence of Santos-Dumont's success, men of letters and science speculated about the future of airships. Most of the prognostications centered on the prospect of air travel. "It will no longer be an absurdity to imagine that air machines may one day compete with electric cars and two-penny tubes to relieve the traffic of great cities," *Westminster Gazette* declared. "Nervous people will naturally conjure up possible terrors of breakdowns and collisions in mid-air; but necessity keeps a motherly eye on invention, and parachutes may be kept on the aerial trains of the future to provide for such contingencies."

Few commentators in the first few years of the twentieth century predicted that flying machines would be used as offensive weapons. To be sure, most of the aeronauts themselves had not given the subject much thought. Since the earliest days of aerostation, various militaries had taken an interest in balloons, not as weapons but as unarmed aerial scouts. In 1794, only a decade after the Montgolfiers had demonstrated the first balloon, the French revolutionary government established a corps of *aérostiers* to serve as the eyes of the infantry. British and American military planners soon followed the French lead in incorporating balloonists into their armies. In the American Civil War, both sides relied on observation balloons to pinpoint enemy positions and assess the damage from ground assaults. In the Franco-Prussian War, in 1870, scores of Parisians actually escaped in balloons from their besieged city.

At various times in the eighteenth and nineteenth centuries, balloonists had volunteered to carry weapons, but military authorities turned them down for good reason: Free balloons, with no means of steering or powering them, were hard enough to guide for observation purposes, never mind the added control that would be required if they had to maneuver above potential targets. In 1793, the Montgolfiers themselves, in an effort to aid the French revolutionary government, offered to drop explosives on the rebellious city of Toulon. In 1846, during the Mexican-American War, the St. Louis balloonist John Wise sent the War Department detailed plans for driving the Mexican army from the Castle of San Juan d'Ulloa in Vera Cruz. He advocated dropping nine tons of bombs from a balloon floating beyond the range of gunfire a mile above the castle. Tethered to a warship by a five-mile cable, the balloon would be reeled in after unloading its charge. But the War Department, Wise said, "was not sufficiently advanced in its ideas to give the proposition the consideration it deserved."

Austrian military planners, on the other hand, were more advanced in their thinking, and in 1849 they authorized the first aerial assault in world history. They instructed a fleet of 124 balloons under the command of Lieutenant Franz Uchatius to drop rudimentary bombs—small cast-iron containers filled with gunpowder—on the city of Venice, which was resisting Austrian rule. But there were no casualties, and probably no damage, as every bomb but one landed in the water, and the one bomb that reached the city exploded prematurely in midair above the Lido. Mindful of the experience in Venice, no military was willing to use a balloon as an offensive weapon until the twentieth century.

Santos-Dumont, although convinced that powered balloons would usher in an age of world peace, was not averse to using them as defensive weapons. From the time of his earliest ascensions, he had seen how ocean water near the coast was fairly transparent from the air. He realized that balloons could be effectively employed to detect approaching submarines and drop explosives on them if they did not retreat. In 1900, he proposed such a scheme to the French military, which showed no interest.

Among the early champions of heavier-than-air machines, Samuel Langley was one of the few who contemplated their use as offensive weapons. In 1896, after his success with unmanned Aerodromes, he believed that the age of flight was imminent, and he argued that equipping planes with guns and bombs would actually promote the cause of peace. The plane, Langley said, will "change the whole conditions of warfare, when each of two opposing hosts will have its every movement known to the other, when no lines of fortification will keep out the foe, and when the difficulties of defending a country against an attacking enemy in the air will be such that we may hope that this will hasten rather than retard the coming of the day when war shall cease."

Langley was convincing. The flying machine "will make armies a jest," Alexander Graham Bell agreed, "and our four-million-dollar prize battle-ship so much worthless junk." Editorial writers across the country started reflecting Langley's views. An essay in *Leslie's Weekly* (July 28, 1896) was typical:

In all great wars hitherto there has been little or no danger to the commanders and generals. Much less to the kings or presidents or bloodthirsty senates or congresses which declare wars. So long as a king may sit in his palace and order his subjects to the carnage and slaughter of the battlefield without danger to himself; so long as members of parliament or congress may sit in leather-bottomed chairs and vote sufficient taxes upon the people to hire a crowd of poor devils at sixteen dollars per month to go out and fight and be shot at—so long, in short, as those who make both the wars and the trouble are perfectly safe from danger themselves, arbitration conventions and peace congresses may sit passing resolutions till doomsday. But the [aerodrome] threatens to change the face of the whole situation. Suppose it to be possible for a foreign battle-ship to sail within two or three hundred miles of our coast and set free an aerodrome loaded with a ton or so of nitroglycerin, and send it flying over Washington town. Would there be so many jingoes ready to plunge this country into a foolish and futile war with England? Would there be any jingoes, or, in the face of such a danger, any Congressmen in Washington at all? Would not duck-hunting or the funeral junket suddenly develop abnormally irresistible charms? I have a fancy that even the insane and mischievous ambition of Kaisers and czars might be rudely checked.

Although Langley may have been the first to claim that the mere existence of a military aircraft would stop wars before they began, he was not the first to argue that the development of a new weapon would bring peace to earth. The men who invented machine guns and high explosives had made the same argument.

Richard Jordan Gatling, who built the first reliable machine gun, was raised on a plantation in Money's Neck, North Carolina. In the 1830s, he invented a series of automated machines for sowing cotton, rice, and wheat. A smallpox epidemic in the 1840s persuaded him to study medicine—he wanted to save mankind from plagues. He completed medical school but for reasons that are not clear, he never became a doctor. Instead he settled in Indiana and resumed work on agricultural technologies. During the Civil War, he applied his technical abilities to armaments and invented the crank-operated Gatling gun, which could fire two hundred rounds per minute. His motivation, he said, was to save lives. "It may be interesting to you to know how I came to invent the gun that bears my name," he later wrote. "In 1861, during the opening events of the war . . . I witnessed almost daily the departure of troops to the front and the return of the wounded, sick and dead. Most of the latter lost their lives, not in battle, but by sickness and sickness incident to service. It occurred to me that if I could invent a machine—a gun—that would by its rapidity of fire enable one man to do as much battle duty as a hundred, that it would to a great extent, supersede the necessity of large armies, and consequently exposure to battle and disease would be greatly diminished." He saw the gun as a defensive weapon. He envisaged that a single soldier equipped with a terrifying machine gun would deter a whole army from approaching. Modern commentators have called Gatling a hypocrite—he was certainly a shady character, offering his guns to Abraham Lincoln while at the same time joining a secret society of Confederate saboteurs—but his words must be treated in the context of their time.

In the *Social History of the Machine Gun,* John Ellis noted that in the nineteenth century, European and American military establishments thought of war as a dignified activity, in which individual soldiers could display valor. Opposing the trend of the Industrial Revolution, in which the machine was seen as the answer to everything, army officers rejected the idea of mechanized killing. "The bulk of these officers came from those landowning classes that had been left behind by the Industrial Revolution," Ellis wrote. "They tried to make the army the last bastion of the life-style that had characterized the preindustrial world." They viewed the bayonet push and the cavalry charge as the supreme moments on the battlefield, moments that embodied "their old beliefs in the centrality of man and the decisiveness of personal courage." Even in 1914 the rifle and bayonet were the weapons of choice. "The behaviour of certain commanders during manoeuvres just before the First World War perfectly summed up the whole military attitude to the new automatic weapons. When asked by keen young subalterns what they should do with the machine guns they replied: 'Take the damned things to a flank and hide them.' " Once the war started, an arms race ensued and each side quickly stocked up on machine guns, but there were still officers who deluded themselves that the old style of war was not over. As late as 1926, British Field Marshal Douglas Haig declared that "aeroplanes and tanks . . . are only accessories to the man and the horse, and I feel sure that as time goes on you will find just as much use for the horse . . . as you have ever done in the past."

The reluctance of European generals before the First World War to deploy machine guns only applied to battlefields on their own continent. When their empires were intent on extending their territory in Africa, they had no reservations about the mechanized slaughter of the large numbers of natives who resisted them. "Without the handful of machine guns," Ellis concluded, "the British South Africa Company might have lost

Rhodesia; Lugard might have been driven out of Uganda and the Germans out of Tanganyika." When the generals in the Great War reluctantly stocked up on machine guns, they did so knowing their efficacy for mass killing.

The machine gun *did* deter aggression, though not on the battlefield, as Gatling hoped, but in an unexpected domain, the workplace. To discourage laborers who might protest their working conditions, American mining companies placed guards with machine guns in conspicuous places. The National Guard too deployed them freely when called by management to intervene in labor disputes. It was this domestic market for his weapons that initially made Gatling a rich man.

Alfred Bernhard Nobel perfected dynamite in 1867, five years after the invention of the machine gun. The architects of the Industrial Revolution needed high-powered explosives to build roads, canals, and mines. Nobel met the demand, amassing a huge fortune in the process, with production of the explosive going from 11 tons in 1867 to 66,500 tons in 1897. His dynamite was central to the building of the Suez Canal. He also sold explosives to the military, but he believed that their destructive power would eventually serve as a deterrent to war. His confidante and former secretary, Baroness Bertha Sophie Felicita von Suttner, was a peace activist of international fame. Her critically acclaimed novel, *Die Waffen Nieder!* ("Lay Down Your Arms!"), appealed to mothers not to send their sons off to war. The novel impressed Leo Tolstoy, and he wrote to her: "The abolition of slavery was preceded by the famous book of a woman, Mrs. Beecher Stowe. God grant that the abolition of war may follow upon yours."

Von Suttner, who organized international peace conferences, had difficulty convincing Nobel of the merits of her cause. "Perhaps my factories will put an end to war even sooner than your congresses," he told her. "On the day when two army camps may mutually annihilate each other in a second, all civilized nations will probably recoil with horror and disband their troops." If only he could invent a more horrific explosive, he thought he would bring about world peace. It is tempting to dismiss Nobel's view as a self-serving excuse to assuage a guilty conscience, but his biographer, Nicholas Halasz, observed that many of his thoughtful contemporaries shared his belief. Von Suttner also tried to convince Theodor Herzl, the founder of Zionism, to support her conferences, and he wrote in his diary: "A man who would discover a terrible explosive would do more for peace than a thousand of its mild apostles."

In April 1888, Nobel had the disquieting experience of reading his own obituary. His brother Ludwig died on the twelfth of the month, and the newspapers mistook one man for the other. Nobel did not like seeing himself described as "a merchant of death" who had become a multimillionaire by inventing one explosive after another, each more devastating than its predecessor. The premature obituary, combined with Von Suttner's gentle but persistent pleading, changed his views. He was old and sick and knew he did not have much time left to affect his legacy. He wanted to be known as a man who had fostered progress in the world, and so he became a patron of scientific discovery. He was friends with Salomon August Andrée, who worked in the Stockholm patent office and had helped him protect the priority of his explosives. Andrée was intent on leading the first expedition to the North Pole but needed substantial funding. Nobel put up half the money and persuaded the king of Sweden to finance the rest. "If Andrée reaches his goal," Nobel said, "even if he reaches but half of it, this will be one of the successes that cause the kind of noise and fermentation which stirs the mind and accomplishes the creation of new ideas and new reforms." Nobel died on December 10, 1896, seven months before Andrée set out for the Arctic Circle and froze to death in the pack ice, not even halfway to his goal.

When Nobel's will was opened, his nephews (he had no direct descendants) were surprised. It began with a peculiar request, addressing his long-held fear of being buried alive: "It is my express will and injunction that my veins shall be opened after my death, and that when this has been done, and competent doctors have noted definite signs of death, my body shall be cremated." On reading further, the nephews learned that they, and Nobel's two lady friends, would inherit nothing. He had left his entire fortune of thirty-three million Swedish kroner to be distributed as annual prizes "to those who, during the preceding year, shall have conferred the greatest benefit on mankind" in the fields of physics, chemistry, medicine/physiology, literature, and, most important, in the promotion of world peace. To the chagrin of his fellow Swedes, Nobel stipulated that the Norwegian parliament would select the winner of the peace prize, giving it to "the person who shall have most or best promoted the fraternity of nations and the abolition or diminution of standing armies and the formation and increase of Peace Conferences." By entrusting the selection to another country, he was underscoring his desire that it truly be an international award. His will established a thirty-year limit on the peace prize "for if in thirty years they have not succeeded in reforming the present system they will infallibly relapse into barbarism."

In 1905, the first recipient of the Nobel Peace Prize was Bertha von Suttner. She believed, Halasz noted, that it might take longer than thirty years for warfare to be banished but she did foresee the end. In 1893, she wrote in her diary:

> The twentieth century will not end without human society's having eliminated war as a legal institution. Writing in my diary, I have the habit of putting an asterisk over an entry on a situation, sinister or threatening, and leaving a few dozen pages blank. Then I write the question: Well, did it actually happen? See* on page—. A reader in the far future may lift from a dusty shelf this volume, and may check on my forecast. Well, how did it turn out? Was I right or wasn't I? He may then note in the margin (I see him doing it): Yes, thank God. Date 19—?

Von Suttner died in 1914, when her dream of world peace seemed more remote than at any time during her life.

The inaccurate forecasts about the life-saving potential of machine guns, high explosives, and military airplanes did not stop future generations of weapons designers from harboring the same conceit that their inventions were so horrific that they would end war once and for all. On August 6, 1945, Luis Alvarez, the Los Alamos physicist who developed a detonator for the atomic bomb, rode in a chase plane behind the B-29 *Enola Gay*. His job was to measure the energy of the explosion when *Enola Gay* dropped Little Boy, the world's first atomic bomb, on Hiroshima. The explosion lit up the sky and rocked the plane. After securing his instruments, Alvarez looked out the window "in vain for the city that had been our target," but all he saw was the mushroom cloud "rising out of a wooded area devoid of population. . . . I thought the bombardier had missed the city by miles." The pilot assured him that the "aiming had been excellent. . . . Hiroshima was destroyed."

On the long flight back to the airbase on Tinian, an island between Guam and Saipan, Alvarez recorded his thoughts about the bombing in a letter to his four-year-old son, which he intended for him to read when he was older:

The story of our mission will probably be well known to everyone by the time you read this, but at the moment only the crews of our three B-29's and the unfortunate residents of the Hiroshima district in Japan are aware of what has happened to aerial warfare. Last week the 20th Air Force, stationed in the Mariana Islands, put over the biggest bombing raid in history, with 6,000 tons of bombs (about 3,000 tons of high explosive). Today, the lead plane of our little formation dropped a single bomb which probably exploded with the force of 15,000 tons of high explosive. That means the days of large bombing raids, with several hundred planes, are finished. A single plane disguised as a friendly transport can now wipe out a city. That means to me that nations will have to get along together in a friendly fashion, or suffer the consequences of sudden sneak attacks which can cripple them overnight.

What regrets I have about being a party to killing and maiming thousands of Japanese civilians this morning was tempered with the hope that this terrible weapon we have created may bring the countries of the world together and prevent further wars. Alfred Nobel thought that his invention of high explosives would have that effect, by making wars too terrible, but unfortunately it had just the opposite reaction. Our new destructive force is so many thousands of times worse that it may realize Nobel's dream.

•

AFTER WINNING the Deutsch prize, Santos-Dumont received thousands of congratulatory letters. Heads of state sent him medals. Fellow inventors—Thomas Edison, Samuel Langley, Guglielmo Marconi—commended his grit and ingenuity. Parisians who benefited from his donation of the prize money scrawled heartfelt thank-you notes. But it was a letter from a childhood friend, Pedro, that meant the most to him. His countryman recalled the games of their youth:

Do you remember the time, my dear Alberto, when we played together "Pigeon flies!"? It came back to me suddenly the day when the news of your success reached Rio. "Man flies!" old fellow! You were right to raise your finger, and you have just proved it by flying round the Eiffel Tower. You were right not to pay the forfeit; it is M. Deutsch who has paid it in your stead. Bravo! You deserve the 100,000 franc prize. They play the old game now more than ever at home; but the name has been changed and the rules modified since October 19, 1901.

*They call it now "Man flies!" and he who does not raise
his finger at the word pays his forfeit.*

When Santos-Dumont read the letter, he was reminded of
why he had taken up ballooning in the first place. It was for
the romance of it, not for a prize marred by petty Aéro Club
politics. He realized that in his struggle to win the Deutsch
prize, he had been hell-bent on achieving speed at the expense
of his "education as an airship captain." He decided that what
he needed most was more practice navigating. "Suppose you
buy a new bicycle or automobile," he wrote,

*You will have a perfect machine in your hand, without
having had any of the labor, the deceptions, the false
starts and recommencements of the inventor and construc-
tor. Yet with all these advantages you will soon find that
possession of the perfected machine does not necessarily
mean that you will go spinning over the highways with it.
You may be so unpracticed that you will fall off the bicycle or blow up the automobile. The machine is all right,
but you must learn to run it.*

For the moment he was content with the state of his machines. His swiftest airship, *No. 6,* was in fine condition. The
day after the prizewinning flight, his chief mechanic tried to
top off the balloon with hydrogen, but found that it would not
take any more. It had not lost any gas whatsoever in its trip
around the Eiffel Tower. "The winning of the Deutsch Prize,"
Santos-Dumont triumphantly noted, "had cost only a few liters
of petroleum." He could take comfort in the fact that even
with his many mishaps, the airship was no less dependable
than the far more evolved automobile. Of the 170 cars entered

in the 1901 race from Paris to Berlin, only 109 were still running after the first day, and of those only 26 reached the finish line. "And of the twenty-six arriving in Berlin," he said, "how many do you imagine made the trip without serious accident? Perhaps none! It is perfectly natural that this should be so. People think nothing of it. But if I break down while in the air, I cannot stop for repairs; I must go on, and the whole world knows it!"

Santos-Dumont had the naive idea that he could fly in solitude, away from spectators, if he moved his operations to Monaco for the winter. There were good reasons to relocate to the Riviera, but seclusion was not to be one of them. The French press was all too happy to follow him to the glamorous principality, as were, the *Herald* noted, "a number of American 'millionaire' owners of yachts." The Bay of Monaco, shielded from wind on three sides by mountains, the heights of Monte Carlo, and the expansive Royal Palace, was an ideally sheltered place to carry out his experiments. The water itself, he hoped, would cushion an unplanned descent, and a rescue boat would never be far behind.

The biggest draw of Monaco, though, was its ruler, his Serene Highness Prince Albert I, who offered to fund his experiments. The prince was a visionary man of science, one of the world's first conservationists. He was sensitive to balancing development in Monaco with the preservation of its mountainous ecosystem, not an easy task in a country no bigger than Manhattan's Central Park. For three decades Prince Albert the Navigator, as he was called, explored the world's oceans, cataloged the life forms that inhabited them, and pondered ways to promote their long-term survival. In the scientific literature his name was attached to the many species of cephalopods that he discovered. In Santos-Dumont he saw a fellow adventurer, and he was eager to help him master the skies. He had invited the Brazilian to Monaco through an intermediary, the Duc de Dino, and Santos-Dumont accepted the invitation. Prince Albert sent him a map of the region and asked him to choose the best spot for his experiments. He selected the Condamine beach, on the west side of the bay. There, to his specifications, the prince's engineers spent three months building a hydrogen-generating plant and an aerodrome that was even larger than the one in Paris. An empty shell of wood and canvas supported by a steel frame, the balloon house was one hundred eighty feet long, thirty-three feet wide, and fifty feet high.

During the days in late October when it was not certain whether the Aéro Club would award him the Deutsch prize, Santos-Dumont had begun negotiating with the prince's representatives. He offered to make daily excursions in *No. 6* whenever the weather permitted. And he promised that before the end of winter he would fly his seventh airship, already under construction, across the Mediterranean from Monaco to Corsica (a distance of one hundred twenty miles) in fewer than four hours. He would land on the north side of Corsica at Calvi, where Christopher Columbus was born. Word of the promise made its way back to Paris in early November and caused a sensation. No one had ever made such a long journey over water, and a four-hour flight would prove the utility of the balloon because mail steamers required half a day to make the same crossing. How, the press wondered, could it have escaped their notice that he had already started work on *No. 7*? And what an airship *No. 7* promised to be—a sleek racing vessel with two fifteen-foot propellers, one in the front, the other in the back, both powered by a single engine.

Santos-Dumont had long been eager to try ballooning over water. Not just for the thrill and novelty of it, but indeed to demonstrate the airship's utility. In the days since he had won the Deutsch prize, the press had debated whether the airship would be put to practical use or whether it was destined to remain a rich man's play toy. Santos-Dumont wanted to prove that the military analysts who had suggested that powered balloons might be of aid to the navy in reconnaissance were right. To serve as an aerial scout, the airship would not even have to ascend much; the scouting role, he noted, could be "performed at the end of its guide-rope comparatively close to the waves and yet high enough to take in a wide view. For this reason I was anxious to do a great deal of guide-roping over the Mediterranean," and, unlike overland flights, there was no danger of the rope becoming tangled in trees, bushes, or buildings. On a low flight, the guide rope would ensure that the airship maintained a stable altitude. If a gust of wind pushed the airship up, the increased weight of the guide rope would cause it to descend to its previous level. By the same token, a draft that suddenly pushed the airship down would plunge more guide rope into the water, thereby lightening the craft and causing it to rise again.

When Santos-Dumont arrived in Monaco in late January 1902, the balloon facility was nearly completed and it passed his exacting inspection. "It had to be solidly constructed," he said, "not to invite the fate of the all-wood aërodrome of the French Maritime Ballooning Station at Toulon, twice wrecked and once all but carried away, like a veritable wooden balloon, by tempests!" Situated on the boulevard de la Condamine, the main road that ran along the beach, the balloon house was an imposing curiosity. Tourists remarked on its two huge doors, now the largest doors in the world, each fifty feet high by 16.5 feet wide and weighing 9,680 pounds. Tiny wheels at the top and bottom of the doors enabled them to be slid open along tracks that protruded from the building. "Their equilibrium was so well calculated that on the day of the inauguration of the aërodrome," said Santos-Dumont, "these giant doors were rolled apart by two little boys of eight and ten years respectively, the young Princes Ruspoli, grandsons of the Duc de Dino, my host at Monte Carlo."

The adjacent hydrogen plant, at the corner of rue Louis and rue Antoinette, was also impressive in its scale. Six metric tons of sulfuric acid, shipped from Marseilles, were stored behind locked doors along with an equal quantity of iron shavings. When the acid and the iron were combined in a large inert vessel, eight cubic meters of hydrogen bubbled off per hour. At that rate, *No. 6* could be inflated in ten hours. On January 22, 1902, the plant was fired up. The inflation process began uneventfully at 7:00 A.M. By mid-morning, though, the principality was in a state of crisis because the iron-rich runoff from the hydrogen factory had, in the words of one official, turned "the blue Mediterranean into the Red Sea." The alarmed official ordered a halt to the inflation, and Monaco's governing council held an emergency session to address the problem. The prince was out of the country, and the council, knowing his highness's passion for preserving ocean life, feared his wrath. But the country's leaders were also men of science. Santos-Dumont explained to them that the runoff was free of acid. He showed them how it passed through three purifying stages before it was released into the bay. The ruddy deposits, he assured them, were nothing other than ordinary rust, anathema to automobiles and airship engines perhaps but innocuous to animals and plants. In fact, he noted, iron was essential to life.

To dramatize the point, he raised a glass full of runoff water, offered a toast to marine creatures big and small, and swallowed the red liquid in one gulp. Olivier Ritt, the governor-general of the principality, not only declared that the hydrogen production could resume, he apologized for his subordinate's hasty order to halt it. Once again Santos-Dumont, by a combination of reason and charm, had gotten himself out of trouble, and he was promising much bigger things than going to Corsica. The island would be merely a short stop on a six-hundred-mile flight to Africa. "It will be my most ambitious effort so far," he declared. "I shall take no one with me on this trip. The Prince of Monaco was very anxious to accompany me as far as Corsica, but I do not care to shoulder such responsibility yet."

A week later, on January 29, Santos-Dumont guided *No. 6* on two flights over the bay. At 10:30 A.M. the police stopped traffic on the boulevard de la Condamine. The doors of the balloon house were ceremoniously slid open, and workmen led the inflated airship out, looking like a float in a parade, with Santos-Dumont waving proudly from the basket. Weighted down by ballast, the airship bobbed awkwardly as the men walked it across the boulevard to the seawall that ran between the sidewalk and the beach. It was then, he wrote in his diary, that he realized "a miscalculation had been made with respect to the site of the aërodrome itself." The problem was the seawall. From the sidewalk it was only waist high, but on the beach side it plunged fifteen to twenty feet down to the lapping surf. The airship had to be lifted over the wall without damaging the propeller or the rudder and then lowered gently onto the beach. The workmen recruited volunteers from the throng of spectators. They were assigned to raise the airship over the wall, while the workmen went down to the beach and positioned themselves to receive the ship. With its nose pointing obliquely downward and the stern grinding uncomfortably along the wall, Santos-Dumont found himself inclined at a steeper angle than he had ever experienced in flight. But the workmen did their job. "They were at last able to catch and right it," he said, "in time to prevent me from being precipitated from the basket."

Once Santos-Dumont was in the air, he began to worry about how he would land. If he descended on the beach, he would have faced the problem of getting *No. 6* back over the high wall. He could not think of how to do this and feared that the only solution was the unacceptable one of deflating the balloon and wasting the gas. The skies were perfectly calm, so he decided to attempt the bold and somewhat risky maneuver of steering directly into the aerodrome without scraping the sides. "Straight as a dart the air-ship sped to the balloon-house," he wrote.

The police of the prince had with difficulty cleared the boulevard between the seawall and the wide-open doors. Assistants and supernumeraries leaned over the wall with outstretched arms waiting for me; below on the beach were others; but this time I did not need them. I slowed the speed of the propeller as I came to them. Just as I was halfway over the seawall, well above them all, I stopped the motor. Carried onwards by the dying momentum, the air-ship glided over their heads on toward the open door. They had grasped my guide-rope, to draw me down, but as I had been coming diagonally, there was no need of it. Now they walked beside the air-ship into the balloon-house, as . . . stable boys grasp the bridle of their racehorse after the course and lead him back in honor to the stable with his jockey in the saddle!

After breaking for lunch, Santos-Dumont went up again at 2:00 P.M. This time he explored the bay for forty-five minutes, and succeeded in maintaining a steady height of about forty feet above the waves. At one point he was so far from shore that the spectators thought he had started for Corsica, but then he looped back and passed over the casino and palace. As he had that morning, he landed by entering the aerodrome "as a needle is threaded by a steady hand." The prince himself, Santos-Dumont said, recognized "that I ought not be obliged to steer so closely on returning from my flights . . . because a side gust of wind might catch me at the critical moment and dash me against a tree or lamp post or telegraph or telephone pole, not to speak of the sharp-cornered buildings on either side of the aerodrome."

The prince offered to tear down the seawall so that Santos-Dumont could land on the beach and walk the airship across the street into the shed. "I will not ask you to do so much," the aeronaut replied. "It will be enough to build a landing stage on the sea side of the wall, at the level of the boulevard." It took the royal workmen twelve days to construct a large wooden platform extending into the bay.

In the middle of the construction, Santos-Dumont had an unexpected visit from Napoleon III's seventy-six-year-old widow, Empress Eugénie, who arrived in a closed carriage from her secluded villa at nearby Cap Martin. No one could remember when she had last appeared in public. Prince Albert had a couple hours' notice of her arrival, and his laborers hastily decorated the interior of the aerodrome with shrubs and flowers. During her tempestuous reign, she had never shown much interest in science or technology, but in Monaco she implored Santos-Dumont to explain every detail of how the airship worked. He also told her of his plans to fly to Corsica and, if all went well, on to Tunis. And if Tunis, she wondered, why not New York? "I think it will be possible to cross the Atlantic in a navigable balloon," he replied. "If I can obtain good hydrogen on the Riviera, I shall be able to carry 250 pounds of petroleum, which would provide ample fuel to keep the motor working for 15 hours. To cross the Atlantic is only a question of multiplying these conditions."

Word spread fast of the empress's presence. Two thousand people assembled outside the balloon shed hoping to see her when she returned to her carriage. The only awkward moment in the visit was due to the presence of a few journalists who followed Santos-Dumont's trials. Eugénie despised newspapermen, and one of the reporters she particularly loathed, Henri Rochefort, happened to be there. The *Daily Express,* which had its own correspondent at the meeting, described the encounter: "The other day M. Henri Rochefort went over to see Santos-Dumont at the balloon shed at Monaco. As he stood there talking to the aeronaut, Eugénie was announced, and met for the first time in her life the man who had done so much to ruin the Empire and who had seldom spared the empress in his writings. Rochefort stood stiff and erect holding his hat in his hand, while the empress very slightly inclined her head. They did not speak." Eugénie wished Santos-Dumont well in his upcoming trials and departed as quietly as she had arrived, covering her face when the cameras clicked as she walked to her carriage.

On February 10, Santos-Dumont christened the new wooden jetty. At 3:00 P.M. the yellow airship darted into the air, pulling a fluttering scarlet banner bearing the initials P.M.N.D.N., an acronym for the first line of the Portuguese poet Camoën's *Lusiad, "Por Mares Nunca D'antes Navega-dos!"* ("Over seas never before sailed!"). Santos-Dumont headed out of the bay into the open Mediterranean. "The guide-rope held me at a steady altitude of about 50 meters above the waves," he said, "as if in some mysterious way its lower end were attached to them. In this way, automatically secure of my altitude, I found the work of aerial navigation had become wonderfully easy. There was no ballast to throw out, nor gas to let out, no shifting of the weights except when I expressly desired to mount or descend. So, with my hand upon the rudder and my eye fixed on the far-off point of Cap Martin, I gave myself up to the pleasure of voyaging above the waves."

Santos-Dumont allowed himself the luxury of leisurely look-ing about. He saw two yachts sailing toward him down the coast. "Their sails were full-bellied," he said. "As I flew on, I heard a faint cheer, and a graceful female figure on the fore-most yacht waved a red foulard. As I turned to answer the politeness, I perceived with some astonishment that we were far apart already." The wind had increased to a squall. He was only a few hundred yards from Cap Martin, where the empress was watching him from the privacy of her veranda. "Porting my helm, I held the rudder tight," he wrote in his journal. "The airship swung around like a boat; then, as the wind sent me flying down the coast, my only work was to maintain the steady course. In scarcely more time than it takes to write this, I was opposite the Bay of Monaco again." He entered the protected harbor, amid a thousand cheers, cut the engine, pulled in the front weight, and let the dying momen-tum of the airship carry it slowly down to the new landing platform. His men grasped the guide rope and, without *No. 6* actually coming to a stop, they walked it over the seawall, across the boulevard de la Condamine, and into the aerodrome. The whole trip took about an hour.

Two days later he was up again at 2:00 P.M., on his fourth trip over the bay. He planned to fly along the coast toward Italy. Yachts lined the intended route. The boats of Gustave Eiffel and *Herald* publisher Gordon Bennett were ready to help if he got into trouble, and so was Prince Albert's steam chaloupe, launched from the royal yacht *Princess Alice*. Two automobiles, a forty-horsepower Mors and a thirty-horsepower Panhard, paced him along the coast. After a few minutes in the air, however, the wind picked up and it started to rain. Santos-Dumont had to end the trip. As he turned the airship around and headed back, the prince decided that he wanted to be the one to grab the guide rope. "Those with the prince," Santos-Dumont said, in his characteristically understated way, "having no experience of its weight and the force with which the air-ship drags it through the water, did not seek to dissuade him." The prince ordered the captain to meet the airship. In front of thousands of his loyal subjects, "instead of catching the heavy, floating cordage as the darting chaloupe passed it, His Highness managed to get struck by it on the right arm, an accident which knocked him fairly to the bottom of the little vessel and produced severe contusions." Santos-Dumont had been in Monaco for less than a month, and for the second time his actions sent the principality's governors, who watched their prince collapse, into a hurried caucus. But contingency planning turned to cheers when the bruised prince stood up and waved feebly to the crowd. A less forgiving man might have withdrawn his patronage of aeronautics but the prince immediately wanted to know how he could further Santos-Dumont's efforts.

On the following day, February 13, it was the Brazilian's turn to get hurt. "It was a glorious day," the *Daily Mail* reported. "The sea and sky were of a perfect Mediterranean blue, except for the Tête du Chien, a rugged rock which frowns above the Principality, where a nasty cloud hovered in ominous fashion." Expectant crowds lined the shore, and the usual boats had gathered, including the prince's. At 2:40 P.M. Santos-Dumont emerged from the shed in his airship.

> He looked pleased and contented, and the crowd gave him a great reception. As soon as the order was given to let go, the flying machine rose rapidly and pointed straight out to sea, but the balloon did not seem to behave so well as usual. It pitched considerably, and more than once the spectators caught their breath and frightened exclamations broke from the anxious onlookers. But the aeronaut kept on his course, and steering to the left, rose till the trailing guide rope was quite twenty feet out of the water. Suddenly a cry of terror burst from the multitude which watched the airship's progress, for a great gust of wind, blown high up from the ugly Tête du Chien, struck the frail vessel with overpowering force. The balloon seemed to stand on its head [actually, on its tail], and it looked as if it would turn over backwards, but M. Santos Dumont did not lose his nerve—he never does. Quick as lightning he opened the valve to let out the gas, and the airship righted a little and the lower end began to empty.

The immediate danger was over. Though the rudder, caught in the envelope, had broken, the descending balloon was drifting toward the Tir aux Pigeons, the royal pigeon-shooting grounds, which were thankfully devoid of trees and other obstructions to landing an airship.

"Then a new fear seized the gazing crowd, for just beneath the Tir aux Pigeons, which M. Santos-Dumont was dangerously near, is a group of sharp-edged rocks, and it looked as if the aeronaut, powerless to alter his course, would be dashed to pieces." But the airship landed just short of the rocks and began to sink. Launches and yachts sped to the spot. "M. Santos-Dumont was soon up to his armpits in the sea.

"Would he be drowned after all? 'Leave the balloon. Never mind the airship,' cried the men who had crowded to the edge of the shooting ground and were leaning over the rail watching the plucky aeronaut battling for his life. But he would not quit the car." He shouted instructions to the nearest boat on how to retrieve the airship. The balloon was half deflated. Before salvage operations could commence, one end buckled, stressing the cords. A few minutes later, at 2:55 P.M., the envelope burst, "and the ragged pieces of silk flapped backwards and forwards in the wind. This was the crowning disaster. The excitement in the bay redoubled, and 'Save him! Save him!' came from a thousand throats. All that remained of the beautiful airship was gradually sinking into the waves, taking the courageous aeronaut with it, when the Prince's launch quickly ran alongside the car, and M. Santos-Dumont, just in the nick of time, clambered out of the sea and over the gunwale of the boat." The remains of the prizewinning airship were easily hauled out of the water with the exception of the motor, which divers subsequently recovered. "When the wreck of the balloon had been towed ashore, the great doors of the now useless Aerodrome were sadly closed."

Santos-Dumont explained to his royal host what had gone wrong. The balloon, it seemed, had been imperfectly inflated and therefore lacked sufficient ascensional force. To provide

additional lift, he had pointed the nose of the balloon diagonally toward the sky, so that the propeller would help push it
upward. The hydrogen had been relatively cool in the shaded
aerodrome but now was heating up in the open sun. "As a
consequence, the hydrogen nearest the silk cover rarefied rapidly," Santos-Dumont said. "The rarefied hydrogen was able
to rush to the highest point, the up-pointing stem. This exaggerated the inclination which I had made purposely." The
balloon pointed higher and higher, until it looked to be almost
perpendicular to the water.

> Before I had time to correct this "rearing up" of my aerial
> steed, many of the diagonal wires had begun to give way,
> as the slanting pressure on them was unusual, and others,
> including those of the rudder, caught in the propeller.
> Should I leave the propeller to grind on the rigging, the
> balloon envelope would be torn the next moment, the gas
> would leave the balloon in a mass, and I would be precip-
> itated into the waves with violence. I stopped the motor. I
> was now in the position of an ordinary balloonist, at the
> mercy of the winds. These were taking me ashore, where
> I would be presently cast upon the telegraph wires, trees,
> and house corners of Monte Carlo. There was but one
> thing to do . . .

and so he released the hydrogen and descended into the sea.

Santos-Dumont now recognized that he had been careless.
Not only had he failed in the preflight inspection to notice that
No. 6 was insufficiently inflated, he had unwittingly risked his
life in the flight the day before. "Looking back over all my
varied experiments," he wrote,

I reflect that one of my greatest dangers passed unperceived even by myself at the end of my most successful flight over the Mediterranean. It was at the time the prince attempted to grasp my guide-rope and was knocked into the bottom of his boat. I had entered the bay after flying homeward up the coast, and they were towing me toward the aero-drome. The air-ship had descended very close to the sur-face of the water, and they were pulling it still lower by means of the guide-rope, until it was not many feet above the smokestack of the steam chaloupe—and that smoke-stack was belching red-hot sparks! Any one of those red-hot sparks might have, ascending, burnt a hole in my balloon, set fire to the balloon, and blown balloon and myself to atoms!

Prince Albert tried to entice Santos-Dumont to remain in Monaco. The Duc de Dino feted the aeronaut and announced the establishment of a fund to offset the considerable expense of rebuilding the airship. At another grand tribute a few days later, the prince himself toasted Santos-Dumont and told him that he should not let his recent mishap deter him from further experiments. Santos-Dumont responded that he had put the accident behind him and was prepared to fly again.

The day after the tribute, he visited the bank where the fund was set up. He did not want to be viewed as a charity case, so he persuaded the bank officer to close the fund and return the money. The press seized on his action and gave it a heroic spin: "The intrepid aeronaut has decided to decline the pecuniary aid which the Duc de Dino and his friends hoped would further his scientific exploits. So great is his devotion to his scheme that he cannot be persuaded to accept financial assistance, which might bring him to the level of the numerous mercenaries who have sought notoriety in the same manner for the purpose of swelling their private pockets."

Santos-Dumont visited the balloon shed one last time. Emmanuel Aimé, who had stood by his side though all his travails in Monaco, wanted to accompany him, but he insisted on going alone. It was late at night, and he climbed over the seawall and walked to the end of the landing platform. He gazed into the choppy water for more than an hour, testily waving away good samaritans who wanted to know if he was all right. Then he returned to the villa and packed his bags. Without saying good-bye to most of his hosts, he boarded a train and headed back to Paris.

[CHAPTER 10]

"AIRSHIP IS USELESS, SAYS LORD KELVIN"

LONDON AND NEW YORK, 1902

•

AFTER HIS mishap in the Bay of Monaco, Santos-Dumont realized that a water landing was not a guarantee that the airship would not be damaged. He decided to resume his experiments on land. In late February, he returned to Paris from Monte Carlo, not with the intention of continuing his work there but of seeing his old friends. On March 4, he headed to London at the invitation of the British aeronauts who were hoping that he would base his operations there. "Mr. Dumont has recovered from his immersion in the Mediterranean," the *Daily Chronicle* reported. "That was merely an incident, though for him a new experience, in the crowded life of an aeronaut. He brushes it aside, and does not propose, at present, to repeat experiments on the seashore. As he says, there is a lack of landing places in the sea."

As he was leaving Paris, the Aéro Club served him with a writ demanding that he remove his balloon shed from their property in Saint-Cloud within twenty-four hours. "It was a rather unusual way of saying good-bye," Santos-Dumont said. As a further slight, the document referred to the street that led to the aerodrome by its old name, rue Deviris, rather than as rue Santos-Dumont, as it had been unanimously renamed by the Saint-Cloud municipal council. Of course, he could not comply with the twenty-four-hour deadline. He sold the "hangar historique" for a thousand francs to a Mr. Glaizot, who took it apart in eight days and reassembled it on the outskirts of Paris as a car garage.

The wreckage of *No. 6* was still in transport from Monaco, and Santos-Dumont scrambled to find it a suitable home. By a happy coincidence, the concert room of London's Crystal Palace was the same length as the shed in Saint-Cloud. An American exposition was set to open there, and the organizers offered to put *No. 6* on display. The Crystal Palace appealed to Santos-Dumont because, like the Eiffel Tower, its construction was an engineering milestone. When the huge glass bubble with iron struts opened in 1851, it enclosed more floor space, one million square feet, than any other structure in the world. Despite *No. 6*'s extensive damage, Santos-Dumont had the idea not just of displaying the airship but of repairing and flying it. A few large pieces of the balloon envelope were still intact and could be stitched together with additional silk. To prevent the recurrence of the problem in Monte Carlo, in which the gas in the under-inflated balloon had rushed into the tip, he planned to partition the balloon with silk walls into three interior compartments. The lightly varnished silk would be permeable to the gas but resistant to it rushing through and suddenly changing the shape of the balloon. All his future airships would have the compartmentalized interior structure.

While he repaired *No. 6*, he was also working on his seventh airship, a faster version of *No. 6*. "My new *No. 7* has forty-five horsepower, being nearly three times as powerful as *No. 6* was," he told his London hosts. "The increased power is secured without a proportional increase in weight owing to improvements in construction. It has cost nearly $5,000, so it is an expensive business when the balloon gets smashed up."

England was not the only place he was thinking of basing his experiments. The other choice was the United States, which he had not yet visited but planned to do within the month. "If I were to choose my nationality," he announced, "I should certainly become either English or American." Already he was playing London and New York City against each other and tantalizing them with the prospects of dramatic, crowd-pleasing experiments. He promised to circle the dome of St. Paul's Cathedral if the English Aero Club built him a suitable hangar on the grounds of the Crystal Palace. Not only was the cathedral a celebrated landmark, it was now a symbol of technological progress because it had just been equipped with blazing electric lights, thanks to a fifty-thousand-dollar gift from J. Pierpont Morgan. For a flight over New York City, he said he intended "to go up the East River, pass under the Brooklyn Bridge, turn, and then come back above it." So far America was offering the most lucrative prizes. The organizers of the 1903 World's Fair in St. Louis had announced a fund of $200,000 for "the world's first aerial tournament." Palmer Bowen, the fair's representative in Paris, had met with Santos-Dumont and promised $100,000 for the successful demonstration of an airship. Santos-Dumont was hoping that the English would come up with a comparable reward. "With the English I am at home," he said. "I am assured no mean jealous feelings will come from the English Aero Club." An English prize, he said, would "bring competition. And I like, when I make my air journeys, to feel the stimulus that comes of a struggle. I care not so much for the money as that a prize offered will bring me English rivals and put me on my mettle. That is the value of competition."

1. A young Santos-Dumont.

Henrique Dumont, his father

3. His favourite photograph of himself, probably taken in 1906, when he was 33.

4. In the company of his eldest brother (second from right), and the three Villares brothers who married three of his sisters.

5. Dining aloft at his "aerial" table.

1898: Santos-Dumont (with a flower in his lapel), age 25, in front of *Brazil*, his only unpowered ~~oon.~~

7. 1898: Standing in the basket of *No. 1*.

8. 1899: *No. 2.*

9. May 11, 1899: *No. 2* crashing into trees at the Jardin d'Acclimation in the Bois.

10. September 19, 1900: Demonstrating *No. 4* for the International Aeronautical Congress at the Paris Exposition.

11. September 1900: Samuel Pierpoint Langley (right), the dean of American science, at a private showing of *No. 4*.

12. October 1901: In the basket of *No. 6*.

13. 1901: Working at his desk in Paris.

14. 1901: His mechanics inspecting the engine of *No. 6*.

15. 1901: *No. 5*.

16. October 19 1901: Rounding the Eiffel Tower in *No. 6* before an enthusiastic crowd.

PROMENADE MATINALE SUR LA JETÉE

17. 18 February 1902: Immortalized at a carnival in Nice and on a post card sold in Monaco.

19. 1903: Launching *No. 9*, the world's first personal flying machine, in front of his apartment on the Champs-Elysées.

20. 1903: Soaring above Paris's rooftops in *No. 9.*

21. June 1903: Nineteen-year-old Aida de Acosta, the first woman to pilot a flying machine, Santos-Dumont's *No. 9.*

22. His female admirer: Lurline Spreckles, daughter of a San Diego sugar magnate.

23. His male admirer: the writer and illustrator George Goursat, better known by his nom de plume Sem.

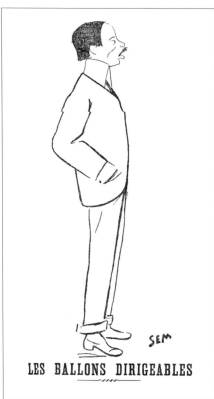

24, 25. Sem's caricatures of his friend.

26, 27. Cartoons about Santos-Dumont appeared in newspapers and magazines around the world.

28. 1906: Testing *No. 14-bis* with a donkey as the power source.

29. 1906: *No. 14-bis, Bird of Prey.*

30. 1909. Transporting *Demoiselle*, the world's first sports plane.

31. 1909: Flying *Demoiselle*.

32. 1909: Sem's caricature of *Demoiselle*.

33. A signed photograph of a monument to his
aeronautical triumphs unveiled in Saint-Cloud on
October 19, 1913.

34, 35. Always well-dressed: Skiing in St.
Moritz after the Great War.

36. A rare smile, in Buenos Aires.

He considered London to be technologically backward compared with Paris. There were automobiles, for sure, but they were not allowed to exceed twelve miles an hour. Cars were still enough of a novelty in 1902 that the local press accounts of his visit always noted the kind in which he rode ("He was driven from Victoria Station to the Carlton Hotel by the Hon. C. S. Rolls in his 20-h.p. Panhard, and his Secretary, M. Emmanuel Aimé, was driven in the electric *coupé* which had been kindly sent to the station by Paris Singer"). In choosing between New York and London, Santos-Dumont was weighing the two cities' emergency services, in case he needed help after an aerial accident. New York hospitals had modern electric-powered ambulances, whereas London did not have a system for rendering prompt aid. An injured person had to make it to the hospital by his own means if he managed to get there at all. To woo Santos-Dumont, London officials promised him that they would introduce regular ambulance service, but neglected to mention that the emergency vehicles would be horse-drawn. There were problems with how the ambulances would be summoned. As one London paper reported, "The suggestion that there might be difficulty in distinguishing fire alarms and ambulance calls is met by the proposal that one ring be given for fire, two if an ambulance."

On April 10, 1902, Santos-Dumont arrived in New York aboard the steamship *Deutschland* and, in the words of the city's tabloids, immediately learned "things which might have shocked one more sensitive to shock." Santos-Dumont, who was accompanied by Aimé and Chapin, his chief mechanic, had taken along a large crate that contained the keel of *No. 7*, but as they disembarked, custom officials seized it. Santos-Dumont told them that he had thoroughly researched U.S.

import regulations. Under the Dingley Tariff Act, he explained, scientists could freely import material for illustrating their lectures and artists could bring their works into the country for purposes of exhibition. He was both a scientist and an artist, he said. But to the officials, he looked like an effete dandy who could afford to pay a stiff tariff, and so they impounded the crate along with other parts of the seventh airship that had arrived a few days before on *L'Aquitane*. They told him that they would charge him 45 percent of the airship's value unless they received contrary instructions from the Treasury Department. The other unwelcome news was that the St. Louis Fair was being postponed a year, to 1904. "The man who flies in the air smiled and shrugged his shoulders at both pieces of news," reported the *Pittsburgh Dispatch*. "Why a gentleman should be invited to come to this country to exhibit the greatest airship ever built and then be made to pay a duty for getting it into the country is something which the philosophy of the Brazilian never contemplated."

Upon his arrival, Santos-Dumont started courting the New York press. He used the words "air port" (perhaps he even coined the words), predicting that New York would become "the great air port of the New World" with a fleet of giant airships bridging the skies from New York to Paris. (The first airport in the New World would be a seaplane port in Tampa Bay that was constructed in 1913.) He said that he expected to fly across the Atlantic in one of his own airships within ten years. The newspapers were less interested in his grand forecasts than they were in the unusual man who was making them. They reported on his way of speaking about his aerial triumphs and mishaps "as calmly as a farmer talks about a sack of potatoes." And they described his appearance:

His eyes are reddish hazel, with an expression of great alertness. They miss nothing. His temples are hollowed and his thin thatch of straight brown hair is slightly grizzled. When one remembers the collision and falls he has survived in his airship, the only surprise is that his hair is not white. The air navigator has a nose of medium height, a trifle bowed, and—marvelous in a man of his pluck and persistence—a chin that distinctly recedes from lip to point. There is something birdlike in his build. . . . His bone structure is of the slightest. He has slender, dainty hands and feet.

The New York papers were engaged in a fierce competition for readers, and the *New-York Mail and Express* had sent a reporter to Santos-Dumont's Paris apartment so it could publish an exclusive story when he arrived in the States. The article provided a surprising picture of the aeronaut's home life:

M. Santos-Dumont, the "King of the Air," the man of ideas, the intrepid aerial navigator, and M. Santos-Dumont, the man at home, are two separate and distinct individuals. One is all enthusiasm, brilliancy, daring; the other is indifferent quite to dullness, and with feminine timidity without the feminine charm. To say which is the real man is difficult, and it is less perplexing to admit at once that he is blessed, or the reverse, with a dual personality. . . .

He has few friends and those he has admit they don't know him well. They admire him because of his daring and ingenuity. But if M. Santos-Dumont lacks the quality of making friends he certainly possesses a power of fascination over the fair sex that neither his appearance nor his manner in society would seem to justify. Women love the mysterious, and Santos-Dumont is a mystery. . . . For the secret may as well out. M. Santos-Dumont does not spend his spare moments in puffing at a cigarette or in sipping innocuous-looking mixed drinks. Far from it. He devotes his time to embroidery, to knitting, and even to the more difficult accomplishment of tapestry making. He revels in all the light bits of needlework that are supposed to belong exclusively to femininity, and, what is more, does not care who knows it. "It is a relaxation," he said when asked about this rather unusual taste; "besides, I like it, and always have."

In his apartment at the Elysée Palace Hotel in Paris, M. Santos-Dumont has "half a dozen embroidered pieces, such as are dear to every housekeeper—tray clothes, tea coverings, cushions, and the like, all showing skill, and a desire for dainty effects." Also two chairs done in a difficult cross stitch of a style only known in France and with colors carefully chosen and pains taken with the design.

But it is knitting to which M. Santos-Dumont is most addicted, and it is some half completed article he instinctively catches up when under any mental excitement or in need of rest. He has the real German trick of clicking his needles in and out, and staring out into space.

In his apartment M. Santos-Dumont lives luxuriously, but it is the luxury of a pampered beauty rather than the gorgeous surroundings of a bachelor of wealth. His suite of three rooms overlooks the Champs Elysées, and is fitted up in a manner to elicit admiration. The drawing room is paneled in white wood picked out in gold, and hung with rose colored silk. The same material hangs below the windows, partly covering much beruffled lace curtains.

The furniture is French and of the Empire period. Gold chairs, delicate of shape, and impossible for use, covered in pale brocades, are scattered about. There are two or three divans with eiderdown cushions, in pale rose and yellows, and one or two charmingly painted screens. In-numerable tables are made an excuse for an infinite variety of bric-a-brac, which is arranged on them with studied confusion.

One corner of the room is devoted to the tea service, and here the host frequently dispenses that social and lady-like beverage. Everything in the room is the best of taste, and nothing in it would for a moment indicate a masculine touch.

The dining room is conventional, hung in tapestries and brightened only by silver sconces on the walls, and a profusion of silver on the buffet.

It is in the bedroom, however, that M. Santos-Dumont has given his taste full sway. Everything is blue and white, and everything is girlish to a degree. Pale blue silk covers the walls, and this in turn has an outer cover of white dotted net. The bed is curtained, and the curtains and the hangings in the windows are fastened back with blue satin ribbons tied in huge bows. On the floors are blue and white rugs. The furniture is white with blue curtains, and against one side of the wall is a large toilet table, all in white, and strewn with a profusion of silver toilet articles.

It is a room that would naturally suggest a young and charming girl, but it is the expression of taste of one of the great inventors of the age, and as such is in keeping with his two-sided character.

In his own attire, too, M. Santos-Dumont shows his fondness for things not masculine. His fingers are covered with rings of various designs, many of them set with colored stones. From his watch fob hang innumerable charms, and his collection of scarfpins is large and varied. A bangle or two invariably can be seen on his wrists, and it is safe to assert that if other jewels were permissible M. Santos-Dumont would wear them. . . .

Society has no charm for him. He is deluged with invitations for dinners, balls, and afternoon affairs. Some he accepts; more often he declines. As a dinner companion he is hopeless, for of small talk he has none. He looks bored and usually seems as frightened as a young girl at her first party. . . .

In some unaccountable way he has acquired the reputation of a lady killer, but there is nothing in his manner to indicate that he is one or is even remotely interested in the fair sex. Women make much of him, but it is difficult for any one to strike an answering chord.

Santos-Dumont was a favorite subject for editorial cartoonists. The Hearst papers ran an illustration of airships circling a mountaintop ski lodge with the caption "Pike's Peak the Future Summer Home of the Wealthy." And a drawing in the *Brooklyn Daily Express* with the caption "Perhaps Santos-Dumont Will Solve the Bridge Problem" showed streetcars suspended from dirigibles floating over the Hudson River. Another illustration showed him gazing at the chest of a zaftig woman and explaining to her that in an emergency he always looked for the softest landing spot.

On his first full day in New York, Santos-Dumont walked around Manhattan and admired the skyscrapers. "These are much taller," he said, "than any building I ever crashed into in Paris." He was disappointed, though, not to see any balloon hangars. He had hoped that aerostation would be a visible part of the city. When he returned that evening to the Hotel Netherland, where he had hung a six-foot model of *No. 6* from the ceiling of his room, the mail had piled up. Most of the letters were from the usual autograph seekers and would-be inventors offering half-baked suggestions, but two piqued his interest: an invitation from Thomas Edison to join him at his home on Sunday and an offer of financial support from a railroad. The Brooklyn Rapid Transit Company, which was in the business of running trains from Manhattan to Brooklyn, wanted to pay him twenty-five thousand dollars to fly from Brighton Beach up the Narrows, around the Statue of Liberty, then up the East River over the Brooklyn Bridge and back again to the starting point. They also proposed a month of flights over Coney Island so that they could offset his fee by drawing additional train riders to Brooklyn. (In hindsight, the interest of the Brooklyn Rapid Transit Company has a certain irony because it was the airplane that would ultimately put many railroads out of business.)

On April 13, Santos-Dumont and Aimé visited Edison in West Orange, New Jersey. For an hour the King of the Air and the Wizard of Menlo Park discussed the state of aeronautics. Their conversation, recorded by a reporter, began on a philosophical note, and Edison did most of the talking.

"I was down in Florida recently," Edison said, "and one day I watched a big bird—I think it was a vulture—that floated about in the air a whole hour without moving its wings perceptibly. When God made that bird He gave it a machine to fly with but He didn't give it much else. He gave the bird a very small brain with which to direct the movement of the machine, but He gave to man a much larger brain in proportion to that of the bird." And so, Edison continued, he had always believed that man possessed the intellect to figure out how to fly. "But you are the only one who has done it," Edison said, nodding approvingly.

"I am sure you have never worked on the problem of aerial

navigation," Santos-Dumont demurred, "or you would have accomplished years ago more than I have done now."

"I don't know about that," said Edison. "I did take it up several years ago and built a specially light motor to be operated by exploding gunpowder. I experimented a lot with it, but I worked with a small model and did not attempt to fly. I gave it all up because I had a number of other things to do which were far more profitable. I'll tell you, if the patent office only protected the inventor sufficiently the problem of aerial navigation would have been solved thirty years ago."

Santos-Dumont looked a bit crestfallen. He turned to Aimé and said that if Edison was correct, man would have flown before he was born.

Noting the discomfort of his guest, Edison quickly added, "But you are all right. You are on the right track. You have made an airship and you have steered it and you have made a step toward the final solution of the problem. Keep at it. But get rid of your balloon. Make it smaller all the time." Edison was suspicious of balloons because he believed that no matter how powerful the motor, they would not survive the battering of a strong wind.

"Have you noticed," said Santos-Dumont, "that I am making the balloon smaller every time I build a new airship?"

"Yes, and that's right," replied Edison, "but make it smaller yet. You are doing well, but it will take a long time to make the thing commercially possible. When you get your balloon part smaller and yet smaller until it is so small that you cannot see it with a microscope then you will have it. Then you will have solved the problem.

"Take the case of the vulture—" Edison continued. As the dean of American invention, he thought it was his role to deliver long soliloquies, and his audiences generally enjoyed every word, even though in Santos-Dumont's case he was speaking to the converted. On this occasion, Edison may have been particularly preachy because of the reporter's presence:

Here is a natural flying machine which is a thousand times as heavy as the air it displaces. In a few seconds of leisurely flight it can sweep over a distance which man finds encumbered with all sorts of obstacles and there is scarcely a flutter of its wings in the operation. There is nothing there but a machine and a small brain and it is not a very reasonable machine either. Why is it that a man cannot make a flying machine as efficient as a bird? A lot of people say that it was never meant that man should fly; that if nature had intended such a thing, man would have been provided with the necessary machinery in his body, such as is now possessed by the bird. But you might just as well say that man was never intended to have any light aside from the sun and the moon and the stars which were originally provided for him, or that he should not move about faster with the aid of wheels because no wheels were supplied to him by nature.

The man or men who really solve the problem of flying through the air will find out nothing new. Powerful motors of wonderful compactness will be applied to a framework of extreme lightness and that will be all there is to it. Doubtless this framework will be something similar to the physical structure of a bird. I do not believe it will be difficult because we have many mechanical devices now which are superior to the devices used by nature in human beings and animals, and I do not see why we may not put together a contrivance which will be at least equal to the machine and brain of the bird.

Edison repeated that he thought Santos-Dumont was on the right track. He agreed that the gasoline engine was the proper power source for a flying machine and said he was sorry his recently invented storage batteries were too heavy to be of help. He explained that he was developing a new compact battery and promised to give the first one to Santos-Dumont for use not as the power source but as an efficient spark generator for igniting the petroleum. Santos-Dumont was delighted with Edison's show of support and asked him if he might resume his experiments in aeronautics.

"No," replied Edison,

> I will not go into anything which cannot be protected from the pirates who live off the work of inventors, and I do not believe it would be possible to secure a patent on either a flying machine or an airship or any part of one that would withstand the test of the courts. If someone should make a commercially successful flying machine dozens would at once copy the models and take away the fruits of the original inventor's labor. There isn't a judge in the country who would hold that there was really any invention in such an apparatus, because so much has been done and written about it that the only difference between the successful machine, which is to be, and the many failures, which have been, will be very slight. I doubt whether any new principle will be discovered on which even a claim for a patent can be made.

Edison's comments were prescient given the long and nasty patent disputes in which the Wright brothers would soon find themselves. Inventors like Edison and the Wrights lived up to the reputation of Americans as cutthroat capitalists.

As Santos-Dumont left Edison's home, the reporter questioned him about their seemingly different approaches to flight. "I was glad to talk with Mr. Edison," he replied.

> He is a practical man. I don't think our ideas are so far apart. He told me I was on the right track. I don't believe in doing away with the balloon entirely yet, but I am making the gas bag smaller and increasing the motive power all the time, so perhaps after a while I shall come to Mr. Edison's plan. Unfortunately, what he says about the reward of the inventor is true, but I have never cared for that part of it. I never tried to get a patent on my part of my airship and never intend to. Whatever money I get in prizes I shall devote to further experiments in the airship line.

Edison was impressed enough with Santos-Dumont's experiments that as soon as the Brazilian left he put in a call to President Theodore Roosevelt. Three days later, on April 16, Santos-Dumont was eating lunch at the White House. "I am happy to see you and to congratulate you," said Roosevelt. "My son is very much interested in your aerial experiments and he hopes you will soon land with your airship in the gardens of the White House."

"I will do my best," replied Santos-Dumont, "and if I succeed I shall be happy to take him on a trip in my airship."

"In that case," said the president, "it is not the boy you will take, but myself."

The tabloids had a lot of fun with the idea of Roosevelt soaring through the air. The *Brooklyn Eagle* ran a cartoon, captioned "The Rough Riders of the Future," showing the president, sword in hand, straddling a balloon like a horse and leading dozens of dirigibles into battle.

Of all the Roosevelts, it was actually the president's daughter, Alice, who seemed to have the best chance of making an ascension with Santos-Dumont. She turned up at his side at a dinner party that the Brazilian embassy threw in his honor. They talked airships, and he told her that it was too great a responsibility for him to take up the president of the United States. She asked him if he intended to fly in New York before the fair in St. Louis. He told her that he was thinking of it. "Then you will take me up in your balloon?" she asked. He thought that she was just making conversation and he jokingly replied that she would be the first woman to ride in one of his airships. But she was quite serious. "You know," she said, "we live on Long Island near where you will sail, and I shall hold you to your promise."

"Very well," he said. He subsequently told friends that if she insisted, he would not back out.

After leaving the White House, he visited Langley at the Smithsonian. Langley showed him the latest model of his flying machine and explained that he was constructing a larger version capable of carrying a man but was having difficulty finding someone to pilot it. "In that case," said Santos-Dumont, "I am at your orders and will try to make it work in the open air."

Santos-Dumont was simply being polite. He did not believe that heavier-than-air flight would actually succeed. "Until some absolutely reliable motor can be found, at once light and powerful, the aeroplane can never be really tested," he told a reporter.

At present there is no motor not liable suddenly to stop, at a critical moment perhaps. With a heavy aeroplane such a *contretemps* means certain death to any one who may have trusted himself to its carrying power. A working model and a practical flying machine capable of sustaining the weight of a man are two very different things. The model may work admirably, do everything that its inventor claims for it, and yet, no sooner is it carried out on a larger scale than all its qualities are seen to have deserted it. It is as if there were some element it was impossible to define which eludes all investigation. So far, in spite of its manifest drawbacks, the "lighter than air" appears to be the direction in which practicable results may be more profitably pursued.

Before leaving Washington, Santos-Dumont visited the Treasury Department and argued for the release of *No. 7* without a tariff on the grounds that his airship was a scientific apparatus. Because no powered balloon had ever flown in the United States, the Treasury officials wanted to know how he could be sure that the machine was science rather than fantasy. Santos-Dumont turned to Aimé and, in earshot of the officials, remarked on the irony of being able to steer the airship around the Eiffel Tower more easily than he could steer it through U.S. customs. Even Langley could not pull strings to free the airship, and Santos-Dumont wished that he had asked President Roosevelt to intervene. The Treasury informed him that the tariff was a steep $630.

Next Santos-Dumont made a quick trip to St. Louis. The organizers of the world's fair had proposed a race between St. Louis and Chicago. While inspecting the site of the fairground, he convinced the organizers that the distance was beyond the reach of any powered balloon. Besides, he said, such a race would not be good for the spectators because the airship would soon be out of sight. Instead he proposed a five-mile triangular course, with colorful captive balloons marking each of the turning points.

Back on the East Coast, Edison called a press conference and announced that he had found Santos-Dumont's work so promising that he was urging the Brazilian to organize the first aeronautics club in the United States. "There are plenty of men to be interested in such an enterprise," Edison said, "and I would probably join myself." This was the kind of public expression of confidence that Santos-Dumont needed, and it came just in time. On April 19, one of the implacable foes of manned flight, Lord Kelvin, arrived by steamship in New York. In his first interview, before he even left the docks, Kelvin denounced the idea of the airship with a vehemence that Santos-Dumont had not been subjected to before.

Under the headline "Airship Is Useless, Says Lord Kelvin," New York tabloids presented the story as high drama. It was the case of a geriatric scientific legend, whose health had suffered on a choppy trip across the Atlantic, summoning whatever strength he had left to expose a charlatan:

Bowed by the weight of four score years, white haired and feeble, Lord Kelvin, foremost scientist of the century, walked ashore yesterday from the Cunarder *Campania* leaning heavily on the arm of his wife. His once stalwart form is beat, his face thin. A thick white beard that sweeps his chest somewhat takes from the sharpness of his features. But the eyes, those eyes that have done so much in peeking into the hidden mysteries of electrical science especially, are as undimmed as ever. . . . On the pier he was assisted to a bench. He sunk heavily into it, as if physically exhausted. But the moment the Santos ship was mentioned his interest was instantly awakened. It was as if a reviving draught had been given to a man about to slip into unconsciousness. His form straightened. He looked up. His eyes took on a look of keen inquiry.

"What of the Santos ship?" he asked. "Ah! You wish my opinion. Well, it is easily given. I think it is of no practical value whatever." His voice was weak and quivering as it began. It gradually grew stronger.

"The airship of Santos-Dumont is a delusion and a snare," continued Lord Kelvin. "The idea of paddling balloons with oars is an old one, and can never be of any practical use. I cannot see how Santos-Dumont can have made such a stir. His plan is useless, useless," and Lord Kelvin shook his head and raised his hands in a deprecating way. . . . "Why, an airship of that type to carry passengers—that is, passengers who would pay to be carried—would not be possible."

Two weeks later, Santos-Dumont left for Europe. Before he departed, he made a point of speaking to reporters from the same pier that Kelvin had: "Lord Kelvin's statement that my airship would never be of any practical use I do not take much stock in, simply because everyone knows he is not an authority on flying machines. I admire Lord Kelvin, but do not like to hear him express opinions about a subject of which he knows practically nothing. In direct contradiction to the scientist's remarks comes the statement of Thomas Edison, who told me he thought I had solved the airship problem." And then he apologized to the press for not having flown in America. "That is the only way, I think, which will stimulate capitalists into building flying machines," he said. "That is the way the automobile was brought out. One was built and shown in Paris, and in a very short time it had been improved on and new ideas developed until we got what we have to-day, every other vehicle in the street a horseless carriage."

He said he was gratified by his reception in America and hoped to return in August but would not commit himself to flying around the Statue of Liberty or over Brighton Beach. He objected to his Brooklyn sponsors' plan to charge admission for the flights and give him a percentage of the gate. "I am an amateur," he said, "and to give exhibitions under the conditions they proposed would savor of professionalism." He revealed that he had sold *No. 6,* which was being repaired in London, to the Brooklyn Rapid Transit Company. "I shall leave my *No. 7* in this country," he said, "and if possible I shall take it to Washington next winter and make some flights there. With this machine, which will carry four persons, I could sail from New York City to Washington without any difficulty at a rate of forty miles an hour. Yet this is only the beginning of what will be realized within a few years if you Americans take hold of the problem." He said he was building another airship for a Chicago man, whose name he would not reveal. "If any one will give me a million dollars to build a machine," he concluded the interview, "I will make an airship that will cross the Atlantic in two days and have a capacity of two hundred passengers."

Even his loyal supporters were now wise to his making grandiose pronouncements that he did not fulfill, and so when he reached Paris, a journalist was waiting for him on the dock. You cannot be serious about the transatlantic flight, the newsman said.

"Certainly I meant it," Santos-Dumont replied, "and no part of the proposition is impossible. As for speed, bear in mind that my latest and previous balloons have travelled faster through the air than the *Deutschland* does on the ocean. . . .

Build an airship on the plan of my existing balloons, but on an enlarged scale, and a speed equal to doing the distance between New York and Paris in two days will be the result."

The journalist pressed him on whether he really needed a million dollars to pull it off.

"As for the necessity for a million dollars, have you considered the cost of two sheds alone, one in America and one here? To doubly house such a balloon properly, with the necessary site and plant, would entail an outlay of $300,000. Don't forget that ballooning of this kind costs a pretty penny. Why, I expended $20,000 on housing one of my comparatively small balloons."

"Who would put up that kind of money?" the man wondered.

"As for capitalists being interested in aeronautics, I found in America men of means—men of the richer class— were exceedingly interested in my experiments. In regard to public timidity about going up in the air, recall the fact that men and women soon become used to changed conditions and journey to-day as they would have feared to do a few years ago. Then you couldn't induce a woman to ride in an automobile. You see them speeding everywhere now. When President Roosevelt's daughter offered recently to take a journey in my airship it was not a joke, but a serious proposition to do so."

"And you really believe transatlantic aerial passages will be accomplished?"

"They certainly will be."

"When the million is forthcoming?"

"When it is forthcoming, it can be done. I shall be ready to try it."

"And suppose you were caught in a storm at sea, such as endangers a steamer."

"The steamer cannot travel with or as fast as the storm, and is endangered. A balloon can do so. Stop the machinery and the balloon will ride on with the storm, 'losing ground,' to be sure, but riding on. Or you can mount higher and ride over the storm. Given a large balloon and proportionate power, and the, to some people, seemingly impossible transatlantic feat can be accomplished."

"For a million?"

"For a million."

Santos-Dumont spent only a few nights in Paris. He was eager to reach London and see the progress his mechanics had made in restoring airship *No. 6*. The English Aero Club had come through by building a balloon shed next to the polo grounds at the Crystal Palace, and he was honor bound to return the favor by making a few flights over the city. He needed to do this soon so that he could ship the balloon off to its new owners on Coney Island. When he arrived at the Crystal Palace, he saw the balloon hanging in the concert hall. Overcome by fond memories of circling the Eiffel Tower, he announced that he would fly it within the week. On the evening of May 27, two of his workers deflated the balloon, wrapped it for protection in a tarp, and carefully transported it from the Crystal Palace to the new shed. Except for short meal breaks, the men never let the folded-up balloon out of their sight—they even slept next to it. Because one side of the shed was still under construction, strangers could have wandered in, but the men saw no suspicious activity. On the following day, Santos-Dumont supervised the balloon's unveiling. He watched his workers place the balloon, still covered by the tarp, in the center of the floor. The corners of the tarp had been knotted together above the balloon so that none of the silk fabric was visible. The men untied the knot and pulled back the protective covering. They saw at once that something was wrong. The silk near the intake valve was ripped. They frantically spread out the balloon and found two series of concentric gashes a third of the way around the balloon. At first Santos-Dumont said nothing; he paced and kicked the ground. Then he sputtered, "It is an outrage. I never expected it here. In Paris I had fears. This has been done with a knife." He proclaimed it the work of a "madman"—a malicious competitor. "They say we aeronauts are all crazy," he said. "Perhaps this accounts for it."

The police had a different explanation. They were convinced that the damage was caused by folds of silk catching in an eyelet of the valve and being further torn by the weight of the balloon. "This cannot be," said Santos-Dumont. "My men have packed the balloon more than twenty times and nothing similar has ever happened before. Occasionally there have been little holes noticed, but never anything so serious as this. I could not trust my life to that balloon. It is beyond repair. When it fell into the sea at Monaco, it was torn clean in half, and could be repaired, but this is impossible."

But the police were not convinced. They called in Stanley Spencer, a well-known British aeronaut, to inspect the damage. He agreed with them that the valve could have precipitated the gashes. The fabric, he said, had deteriorated "by the action of the gas used in previous ascents and by the effect of the heated atmosphere and the sun's rays while the vessel remained on exhibition in the Palace buildings." He added that he was "not surprised at the accident having occurred, for as a rule the life of a balloon was limited to two seasons, and the somewhat severe vicissitudes to which the vessel had been subjected, notably its fall into the sea at Monaco, had no doubt weakened it considerably."

Santos-Dumont denounced Spencer's explanation as absurd and hastily departed for Paris on July 4, leaving the wounded gasbag behind so that it could be displayed to the public for a couple of days. "When we first noticed the damage," he said, "we could see it had been done by someone with a sharp instrument. The holes were behind each other, fold on fold. We could see them going through and through. Nothing would make me think otherwise. It must have been done out of malice."

He still had not forgiven the Paris Aéro Club, so he now focused his attention on New York. Gotham was looking better now that a bona fide Aero Club of America had been formed with Thomas Edison, Alexander Graham Bell, and Nikola Tesla among its charter members. Supported by the Brooklyn Rapid Transit Company, which still hoped that Santos-Dumont would fly over Coney Island, the club was building him a mammoth balloon shed—175 feet long, 125 feet wide, and 60 feet high—on the boardwalk next to the Brighton Beach Theater. Aside from housing a hydrogen generator and a machine shop, the shed included an apartment for Santos-Dumont and his assistants. Mindful of the vandalism in London, the transit company promised that "a score of private watchmen will be on guard night and day to prevent annoyance by cranks or others."

Santos-Dumont was pleased with the progress in New York but still called for a prize. "I don't intend to prepare my machine and fly about New York in an aimless manner, simply to demonstrate that I can navigate the air," he said. "I want a definite task put before me. My workmen in New York will have my balloon ready when I land if a proper inducement is offered, as I will go to work immediately, in all probability at Coney Island."

On July 12, while Santos-Dumont was still in Europe, pieces of *No. 6* arrived at Brighton Beach in crates and oil-cloths, and five workers from France assembled it under the watchful eyes of railroad security guards who kept spectators away. George Francis Kerr, the secretary of the Aero Club of America, received a telegram from the aeronaut: "Am sailing 17th. Do my men want anything from here? Cable Santos. Ritz Hotel." Kerr replied that his men wanted French ciga rettes. On July 20, the airship, practically identical to the craft that circled the Eiffel Tower but with a new balloon, was inflated for the first time, and Kerr displayed it to city officials. The balloon had the familiar shape of a cigar with pointed ends and measured 115 feet in length and 19 feet in diameter at the widest. The balloon was ballasted with fifty sandbags swung at equal distances from one another along the sides. The officials were impressed by the "delicate workmanship" of the sixty-five-foot frame. "Nothing but the most highly sea-soned cypress is used in its construction, and the pieces are so nicely put together that the connections are hardly discern-ible."

Two days later Santos-Dumont arrived in New York on the steamer *Kronprinz Wilhelm*. For a moment his acquaintances did not recognize him because he had shaved off his mustache and shed his jewelry and hat. He checked in to the Waldorf-Astoria and had lunch with Kerr, who presented an offer from the Aero Club of America of twenty-five thousand dollars if he successfully circled the Statue of Liberty from Brighton Beach.

Santos-Dumont seemed pleased. Not only was the money good but the source of it, Edison and Bell's organization, guaranteed him respectability.

That afternoon he visited the Brooklyn shed. After greeting his workmen and giving them their cigarettes, he inspected the airship, "as a horseman would inspect a horse or a sea captain would look over a vessel. . . . He tried the thin bands of steel wires that connect the balloon with the delicate framework beneath it. Then he tested the solidity of the motor in the centre of the machine. Next he tried the wicker basket. He pulled at it and he pushed at it, evidently with the purpose in view of learning just how solidly it was attached to the framework."

He seemed satisfied with the condition of the airship but rather than setting a date for a flight, or negotiating details of the prize with the aero club, he went to Newport, Rhode Island, for a week of parties and banquets with the New York society set who spent the summer there. The Newport casino had just been illuminated for the first time with electric lights, two thousand red, white, and green bulbs, and Santos-Dumont joined Alice Roosevelt to view them. He did not wager any money at the gaming tables, however, because he believed that gambling, like smoking, was immoral.

Rhode Island newsmen were disappointed that he did not fly over Newport, and so they trumpeted an inconsequential story of his courage on the ground:

Newport, R.I., Thursday—M. Santos-Dumont evidently knows something about flying horses as well as flying machines. The celebrated aeronaut was the guest this morning of Mr. W. Gould Brokaw in an automobile run down Bellevue Avenue, and under the leadership of Mr. Brokaw he assisted in subduing a horse that had been made frantic by the sight of Mr. Brokaw's machine. Mr. Lewis Coelman Hall was also of the party.

Seeing that the horse, which was being driven by a woman, whose name could not be learned, was about to pull his fair driver from her seat, all three men . . . left the "auto" and tackled the bucking horse in true cowboy fashion. They were not long in quieting the animal, and though at one time when the horse was trying to clear himself from the shafts, the woman seemed in great danger, nobody was hurt, and no real damage was done.

M. Dumont visited the Casino this morning and naturally attracted much attention. He met Mrs. Stuyvesant Fish and Mrs. Oliver Harriman, and for ten minutes the party was absorbed in flying machines, or "air ships," as Mr. Dumont prefers to call them. He leaves town to-morrow, but will return in a few days.

Back in Brooklyn, all was not going smoothly with the airship. "Two hundred people were badly frightened," the *Herald* reported, when Santos-Dumont's workmen tested the propellers on the afternoon of August 10. They were revolving with such force that several women behind a fence got scared "and tried to push their way out of the crowd. Those in the rear tried to get nearer the airship. The fence gave way and fell with a crash, striking the big propellers as it fell. In a second there was much excitement in the building. The propellers flew around with lightning velocity, and several persons narrowly escaped being struck." Upon inspection, one propeller blade was found to be slightly damaged but nothing that could not be easily repaired.

An accident the next day was more serious. After the workmen started the engine, they heard "a loud report followed by a tremendous crash. . . . The steel bands that form the frame for the silk balloon were bent and broken, while the silk itself was torn into shreds. The propeller was also twisted out of shape." The men were not sure what went wrong, but the fact that the machine had collapsed while still on the ground, protected from the wind by the shed, obviously did not bode well.

When Santos-Dumont heard about the incidents, he did not appear concerned. He shrugged them off and told representatives of the aero club that he would circle the Statue of Liberty within the week. But he asked them to keep his plan secret because he did not want to disappoint anyone if he had to cancel the flight at the last minute on account of nasty weather. On August 14, when his friends in the city thought he was spending the day with the Brazilian minister, Santos-Dumont calmly walked aboard the steamship *Touraine* and sailed for France. Although his name did not appear on the passenger list, which was routinely made public before the voyage, deluxe cabin number 333 had been reserved for him.

"I am disgusted with it all," he said as he departed,

and I'm returning to Paris until the St. Louis Exposition, when I shall positively fly. I have no doubt that the American public will be disappointed at my going away without accomplishing the trip from Brighton Beach to the Battery, but the . . . prize which I understood had been offered was so long in forth coming that I grew disgusted. I fully intend to make the Battery trip, and when I return perhaps shall be able to do so. I am not wealthy and I cannot make an exhibition of that kind without some assurances that my expenses will be paid.

When George Francis Kerr was told of the aeronaut's departure, he refused at first to believe it. The prize money was secure, he said, and nearly all the preparations for the flight had been made. "It was only yesterday that Mr. Dumont told me to order the hydrogen gas with which his balloon was to be inflated," Kerr said. "And he, at all times, seemed to be most enthusiastic over the trip. If Mr. Dumont had not made other ascensions I might explain his hurried departure by a loss of nerve, but, as everybody knows, he is a brave man, and has made several exhibitions to prove this. . . . Of course, all we can do is wait until Mr. Dumont returns or arrives on the other side, and then perhaps we shall obtain a more lucid explanation of his now unexplainable conduct."

But a satisfactory explanation was never forthcoming. In fact, his remarks in Paris were stranger and more strident. Describing the alleged failure of the Aero Club of America to produce the twenty-five thousand dollars as "the greatest disappointment of my life," he said that he had given up the hope that America was bold enough to take the lead in the development of "the romantic science of air locomotion." He chastised himself for not recognizing

in the first place that France is the only nation possessing sufficient imagination and faith to enter seriously and confidently into such a field. My experience in England and the United States proves that the Anglo-Saxons have not the necessary temperament, and that therefore France must remain the scene of the aeronaut's future struggle with the air until the problem is solved. New Yorkers will be willing to put up money only when millionaires promise to take up airshipping as a fad. Chicago's talk has proved to be merely hot air. In short, America's entire attitude toward the matter is one pure, gigantic bluff. I lost time and furnished food for the comic papers. That was all my trip to the United States amounted to.

THAT SUMMER Santos-Dumont's mother, whom he had rarely seen in the past decade, took her own life in Portugal, where she had moved as a widow to be close to her daughters. The circumstances of the suicide are not known because her children made it appear that she had died of natural causes. Santos-Dumont, as the son who lived nearest to her, had the responsibility of retrieving the corpse and eventually reuniting it with his father's body in Brazil.

•

SANTOS-DUMONT CONFIDED in Sem that he had lost his nerve. He feared that he would have an accident before achieving his aeronautical dream—a personal flying machine that was as dependable as the automobile. In his lifetime, he wanted to be able to fly on the spur of the moment, to visit friends or go to dinner. The airship he now possessed was a mercurial contraption that could ascend smoothly one day and lose its whirling propeller the next. In late 1902, he finally designed the machine of his dreams, airship *No. 9, Baladeuse*. He also continued to work on *No. 7*, the racing airship for St. Louis. (There was no *No. 8* because he regarded the number as unlucky after his crash at the Trocadéro on the eighth of the month.)

In 1902, he also had to find a new home for his airships since the Paris Aéro Club had evicted him from the Parc d'Aérostation. After months of negotiation with city officials, who convinced potential new neighbors that he was not going to crash through their roofs, he settled on the suburb of Neuilly St. James, next to the Bois. The site was a large vacant lot surrounded by a high stone wall, which gave him the seclusion he now coveted. His workmen would be the only witnesses to any setbacks on the ground. The wall kept passersby from happening on one of his propeller blades careening around the hangar like an out-of-control guillotine. The new facility, which he called the first of the world's airship stations, was not completed until 1903. It had a workshop, of course, a hydrogen-generating plant, and living quarters where he and his workmen could rest before a dawn flight. There were seven hangars in all, the largest being 170 feet long, 32 feet wide, and 45 feet high. He employed fifteen men at Neuilly St. James. While the wall afforded him some privacy, curiosity seekers often scaled it to get a glimpse of France's most famous aeronaut.

In the spring of 1903, he started flying *Baladeuse,* "the smallest of possible dirigibles," and it lived up to his expectations. With a volume of 7,700 cubic feet, *No. 9*'s balloon was a third the size of *No. 6*'s. Its Clément motor produced only three horsepower; at twenty-six pounds the motor was so light that the whole airship was small enough to use instead of his car. That summer *Baladeuse* was his mode of transportation. He went shopping in *Baladeuse,* visited friends, and regularly flew to restaurants and clubs where he would hand the doormen the reins of his aerial steed.

To celebrate the success of *Baladeuse,* he organized one of his signature aerial dinners at the Elysée Palace Hotel. According to Minnie Methot, a socialite from New York, Santos-Dumont ushered his guests into a banquet hall filled with seven-foot-tall tables. The chairs, correspondingly high, were mounted from portable steps. The waiters walked about on stilts. A miniature airship, which was suspended from the ceiling, rapidly circled overhead. After the meal, everyone adjourned to ten-foot-tall chairs in a neighboring room to witness a game of billiards played on a ten-foot-high table. "The sensation is novel and exhilarating," Methot said, "and it was told to us that evening that Santos-Dumont has in his experiments become so thoroughly imbued with aerostatics that he could no longer dine in comfort at the ordinary altitude, but preferred the aerial table for daily use."

In *Baladeuse,* another observer noted, Santos-Dumont

> sought to prove to an unbelieving world the practicality of lighter-than-air craft. At a height of sixty-six feet above the ground, a level he maintained by means of an enormous drag rope of some 132 feet in length, the young adventurer scientist would float above the populace, over the boulevards, descending now and again for an aperitif while he dropped anchor at a café, and again watching the antics of the crowd as he leaned over the edge of the basket. Never mind that the drag rope caught in trees and buggy tops, flicked the backs of terrified horses and enraged small dogs as it swayed through the streets like an enormous snake on end. Dramatic in the extreme, this performance only heightened his popularity.

Some people, like André Fagel, encountered the aeronaut day after day:

> I had just sat down on the terrace of a café and was enjoying an iced orangeade. All of a sudden I was shaken with surprise on seeing an airship come down right in front of me. The guide-rope coiled around the legs of my chair. The airship was just above my knees and Mr. Santos-Dumont got out. Whole crowds of people rush forward and widely acclaim the great Brazilian aviator; they like courage and sportsmanship. Mr. Santos-Dumont asks me

to excuse him for having startled me. He then called for a drink, got on board of his airship again and went gliding off into space. I am glad that with my own eyes I have been able to contemplate the flying man.

The next day I went to the Bois de Boulogne. Just as my car was about to go through the Porte Dauphine, the flying man landed in the roadway. The policemen rushed forward, stopping the passers on foot, on horseback, and in all kinds of vehicles. For a few minutes all traffic was held up as far as the Arc de Triomphe. The trotting horses snorted, the motors loudly hummed on the cars being brought to a sudden stop, shaking up the occupants. The nursemaids taking children out for an airing in the Champs-Élysées became nervous. What was the matter? Was it a riot? Had the King of England returned to France? No. It was Mr. Santos-Dumont on another of his aerial promenades.

On June 23, 1903, Santos-Dumont decided to land *No. 9* in front of his home for the first time. The police discouraged him from descending on the Champs-Elysées, one of the city's busiest streets. He did not want to cause a traffic jam, and so he set out before sunrise:

Knowing that the feat must be accomplished at an hour when the imposing pleasure-promenade of Paris would be least encumbered, I had instructed my men to sleep through the early part of the night in the air-ship station at Neuilly Saint James, so as to be able to have the "No. 9" ready for an early start at dawn. I myself rose at 2 A.M., and in my handy electric automobile arrived at the station while it was yet dark. The men still slept. I climbed the wall, waked them, and succeeded in quitting the earth on my first diagonally upward course over the wall and above the river Seine before the day had broken. Turning to the left, I made my way across the Bois, picking out the open spaces so as to guide-rope as much as possible.

When I came to trees, I jumped over them. So, navigating through the cool air of the delicious dawn, I reached the Porte Dauphine and the beginning of the broad Avenue du Bois de Boulogne, which leads directly to the Arc de Triomphe. This carriage promenade of Tout-Paris was absolutely empty.

"I will guide-rope up the Avenue of the Bois!" I said to myself gleefully.

What this means you will perceive when I recall that my guide-rope's length is barely 132 feet, and that one guide-rope's best with 66 feet of it trailing along the ground. Thus I went lower than the roofs of many houses on each side. . . .

I might have guide-roped under the Arc de Triomphe had I thought myself worthy. Instead, I rounded the national monument, to the right, as the law directs. Naturally I had intended to go on straight down the Avenue des Champs Élysées; but here I met a difficulty. All the avenues meeting at the great "Star" look alike from the air-ship. Also, they look narrow. I was surprised and confused for a moment, and it was only by looking back, to note the situation of the arch, that I could find my avenue.

Like that of the Bois, it was deserted. Far down its length I saw a solitary cab. As I guide-roped along it to my house at the corner of the Rue Washington, I thought of the time, sure to come, when the owners of handy little air-ships will not be obliged to land in the street, but will have their guide-ropes caught by their domestics on their own roof-gardens. But such roof-gardens must be broad and unencumbered.

So I reached my corner, to which I pointed my stem

slightly and descended very gently. Two servants caught, steadied, and held the air-ship while I mounted to my apartment for a cup of coffee. From my round bay-window at the corner I looked down upon the air-ship. Were I to receive the municipal permission, it would not be difficult to build an ornamental landing stage out from that window!

The successful flight gave him the courage to execute two aviation firsts with *No. 9*. On June 26, he descended at a children's fair in the Bois. Many youngsters begged him to take them aloft, but their parents forbade it. One boy, though, seven-year-old Clarkson Potter, was so persistent that his parents acquiesced. Santos-Dumont estimated the child's weight and decided that he was light enough to accompany him into the air a few feet. "The boy will surely make an airship captain, if he puts his mind to it," Santos-Dumont said. Potter became the first child to realize the dream of a manned flight, albeit a short, low one.

The second "first" involved a beautiful Cuban debutante, Aida de Acosta, the nineteen-year-old daughter of a prominent New York family. Having just completed school, Acosta had come from America with a few classmates to enjoy the City of Light. A mutual friend brought her to the airship hangar and introduced her to the famous aeronaut. Young society women were expected to be demure, but with her parents absent she "confessed an extraordinary desire to navigate the airship."

"You mean that you would have the courage to be taken up in the free balloon, with no one holding its guide-rope?" asked Santos-Dumont. "Mademoiselle, I thank you for the confidence!"

"Oh, no," she said. "I do not want to be taken up. I want to go up alone and navigate it free, as you do."

He was impressed by her resolve, not to mention her beauty, and after three lessons on the ground and one of his aerial dinner parties, he felt she was ready to ascend the next time the skies were calm. The details of her subsequent mile-long flight, although witnessed by much of Paris, would be lost if it were not for her confession thirty years later. When she descended triumphantly in 1903 as the first woman to pilot a flying machine, her mortified parents begged Santos-Dumont to keep her name out of the papers. They were of the opinion that there were only two appropriate times for a woman to be mentioned in the press, when she married and when she died. Because he had spoken to newsmen before her parents silenced him, he was only partially successful in squashing the media attention. In his memoirs he devoted only a few sentences to the historic flight and did not mention Acosta by name.

Her parents threatened to cut off her inheritance if she pulled any more aerial stunts. She stayed on the ground, and although she kept the *Baladeuse* flight a secret, she did not conceal her fascination with flying machines and the men who piloted them. Later she befriended Charles Lindbergh and married his lawyer, Colonel Henry Breckinridge, who served as the assistant secretary of war under Woodrow Wilson. In the early 1930s, she and her husband were entertaining a young naval officer, Lieutenant George Calnan, at a dinner party in their New York apartment. The men started talking aviation and Calnan explained that he wanted to work on the navy's dirigibles. Not wanting to exclude his hostess from the conversation, Calnan started to explain to Mrs. Breckinridge the rudiments of lighter-than-air flight.

"I've flown dirigibles myself," she interrupted. "They are a lot of fun." Her husband was more surprised than the naval officer—she had never told him of her aerial exploits. Characterizing the historic flight as a "schoolgirl prank," she explained how Santos-Dumont had trained her:

He showed me how to steer the big rudder, how to shift ballast and drop weights, and how to work the propellers. There were three gears for three different kinds of speed, slow, medium, and fast. You worked them by pulling out just one lever. Even when the great day came, I did not take him seriously when he said, "You are to fly over the Bois today." But just before I took off, he worked out a code of signals for me with a handkerchief, telling me that he would be on the ground below at all times during the flight. "Watch my signals," he said. "I will follow you with a bicycle. When I wave left, steer left. When I circle the handkerchief, let the prop go as fast as it can. If I drop it, descend gently." Then, looking very solemn, he tied a cord connected with the gas valve to my wrist. "If you get too high in the air and get frightened, pull that cord," he said. "It will let the gas out of the bag and then you will descend. If you faint, your weight will come down with something of a crash, but it will not kill you."

Then I was ready. They started the engine. With my hand on the steering wheel, which looked just like the big steering wheel of an automobile of that day, and my eyes on the dial in front, I soared out of the hangar and upward. I remember passing over the Café Madri. My little petrol engine was working smartly and smoothly, but it made a terrible noise.

Santos-Dumont was trailing her on a girl's bicycle, which he preferred to the boy's variety because he could dismount without catching his opera cloak on the middle bar. "He worked harder on his pedals than I did on my airship," she recalled. "But never once did I need any help. The ship handled beautifully, flying at constant elevation." When she crossed a field, they became separated because he could not scale a tall fence. For a while she could not see him but remained calm, taking in the sights of Paris. Eventually she spotted him in the distance frantically waving his handkerchief, signaling her to land.

In front of her, beyond a stand of poplars, was the polo field at Bagatelle, at the northern end of the Bois, where a match was about to start between the Americans and the British. From the air she enjoyed a unique view of the high-toned crowd—the bright blazers, the straw hats, the long ruffled gowns, the pastel parasols. In his choice of landing spots, she observed, Santos-Dumont always exhibited the Latin flair for showmanship. The crowd by now had noticed the flying machine. "Santos! Santos!" they shouted as the balloon motored closer. But the figure in the balloon was not as svelte as Santos-Dumont, and was wearing a broader hat than his *chapeau melon. "C'est une Mademoiselle!"* someone shouted, and the spectators rushed forward for a better look at the corsetted figure in a large, fluttering black hat festooned with pink roses. The engine was making a racket, and the polo horses bolted. No one pursued the fleeing animals because they were spellbound by the airborne spectacle—an event so rare that there is no record of a woman ever flying solo in a dirigible again.

"I will never forget how those people gazed at me as I pulled the valve cord to release the hydrogen gas and started to descend," she recalled. "But the question utmost in my mind was how I was to get out of that terrible basket. You see, Santos weighed only 110 pounds while I weighed 130, and while he could shift about easily in the basket, I was firmly wedged in and could not move. In fact, they had to lighten the ship before I started by throwing away some of the sand ballast and removing the searchlight." It was not just the extra twenty pounds that made the basket tight. It was also her fetching, if bulky, Victorian outfit, with the wide bustle, petticoat, flounce, and fichu. Her dapper mentor was known for the inappropriateness of his aeronautical wardrobe, yet she had outdone him.

"In my long black and white foulard with its ruffled, tight skirt the question of alighting before all those men became an embarrassing one," she said. "Finally, by myself, I worked my way out as far as I could and then the six ground-crew men gallantly tipped the basket over on its side and helped me out."

Santos-Dumont came bicycling through the crowd and happily declared that she was *"la premiere aero-chauffeuse du monde."* After helping her adjust her coiffure and hat, she climbed back in the basket and, over the objections of the crowd—*"C'est fou!"* ("It is madness!")—she took off again and returned to Neuilly St. James for a smooth landing. The triumphant round-trip flight lasted an hour and a half.

Although the two of them did not stay in contact, Santos-Dumont kept a picture of the Cuban beauty on his desk in Paris, leaving visitors with the false impression that the two were intimately involved. Brazilian biographies tend to portray a love affair between them, although in fact she was never alone with him. After he died, she told inquiring obituary writers that she hardly knew the man. She saw him only a half-dozen times, and he was too shy to make conversation. The only words he could get out of his mouth, she said, were the instructions for flying *No. 9,* and even those he said haltingly.

Not everyone was pleased by Acosta's flight. "A well known actress of light comedy, who as long ago as two years began begging and imploring M. Santos-Dumont to take her up in one of his airships, is now gnashing her teeth with envy," London's *Daily Telegraph* reported.

On July 11, Santos-Dumont traveled in *No. 9* to meet friends for lunch at La Grande Cascade. Some army generals who were in the park planning Bastille Day military exercises were intrigued with the egg-shaped airship on the lawn in front of the restaurant. They interrupted Santos-Dumont during dessert and requested that he surprise the troops by flying above them during the exercises. He replied that he could not guarantee it because, with only a three-horsepower engine, *No. 9* would not be able to fly in a breeze. Come if you can, the generals told him.

On July 14, the skies were still. Santos-Dumont ascended at 8:30 A.M. and found a stable altitude at three hundred fifty feet. The president of the republic, Emile Loubet, was reviewing soldiers in formation at Longchamp. Suddenly shots rang out, and the president ducked, fearing an assassination attempt. His security detail roused him in time to see the petite figure of Santos-Dumont soaring overhead, firing a twenty-one-gun salute of blank revolver cartridges. The military men were impressed enough to visit him afterward and persuade him to make his airships available to the French government in time of war. He consented, provided that the conflict was not with the Americas.

Santos-Dumont had actually been late for his rendezvous with the troops at Longchamp. He had always had difficulty timing his own flights. His hands were so busy with the numerous cords and controls that he could not pause to take out his pocket watch. Since the embarrassing experience in 1901 of having to ask whether he had made it around the Eiffel Tower in thirty minutes, he had been complaining to his friends that someone should design an aeronautically friendly timepiece. In 1903 or 1904 Louis Cartier, whose grandfather had founded Maison Cartier half a century earlier to provide jewelry to Europe's royal families, addressed the problem. He made Santos-Dumont a wristwatch with a distinctive square bezel and leather band. Brazilian accounts of the aeronaut's life often claim that he oversaw every detail of the timepiece's creation and that it was the first wristwatch in the world, but that overstates the case.

The wristwatch actually dates back to the late 1500s—Queen Elizabeth I sported one—but they were rare for the next three hundred years. In fin de siècle France, the wristwatch was worn exclusively by women and not as a practical timepiece but as a flashy accessory to call attention to slender wrists and pale arms. The first men to wear them were army officers. In the heat of battle, they too could not afford to pause and pull out pocket watches. German commanders strapped special-issue timepieces to their wrists during the Franco-Prussian War and British officers did the same during the Boer War. Santos-Dumont may well have been the first male civilian to have a wristwatch. There is no doubt that he is the one who made it acceptable for men to wear them. Like the other items he wore, wristwatches became necessities for fashionable men throughout the city.

Sem was one of the people who copied his wardrobe. The two lifelong bachelors were virtually inseparable for a few months in 1903. They wore the same suits, same high collars, same hats. They strolled together on the Champs-Elysées and through the Bois. They pretended that they had disguised themselves by donning dark glasses, but everyone of course knew who they were. They had lunch together at La Grande Cascade and supper at Maxim's. Sem drew his illustrations for

Paris's leading periodicals from the aeronaut's workshop and apartment. People speculated that the two men were lovers, and if they were, they were discreet—at formal functions, Santos-Dumont always had an attractive woman by his side. Rumors also circulated that he was involved with one of his mechanics (and later, near the end of his life, with one of his nephews). Santos-Dumont chose to ignore convention with his foppish garb, but he knew that eccentricity and genius were linked in the popular imagination. He also knew how far he could go without losing his fans. He was preoccupied with what people thought of him, and he would not have jeopardized his reputation by admitting to being homosexual.

Although Paris intellectuals tolerated—and occasionally even indulged—homosexuality, the French public was less accepting, and Brazilian society was unabashedly hostile. (The Brazilian newspapers were so conservative that they rarely ran descriptions of his refined appearance.) Even Marcel Proust, who featured same-sex relationships in his novels, wanted French readers to think that he was heterosexual. When a review of his first book, *Pleasures and Days,* insinuated that he was "an invert," twenty-five-year-old Proust denied the slur and challenged the reviewer to a duel. (Pistols were drawn at Meudon in February 1897: Proust was a bad shot, and fortunately Jean Lorrain also fired wide or else the world might have been deprived of *Remembrance of Things Past*.) Homosexual men were accepted in France and Britain as long as they did not flaunt their desires. Oscar Wilde's trial was a cautionary tale of how someone's life could be destroyed if he paraded his homosexuality.

It is not known whether Santos-Dumont and Proust, two years his junior, ever met. They certainly frequented the same places, and Proust's erotic life was intertwined with the automobile and the airplane. In 1908, he fell in love with his chauffeur, Alfred Agostinelli, and led "the life of a cannon-ball in flight" as he was driven around by Agostinelli in the seaside town of Cabourg in Normandy. The two men soon parted company but reunited in 1913 when Agostinelli showed up destitute with his wife on Proust's doorstep. Proust took in the two of them and hired Agostinelli as his personal secretary. Proust hated the wife, and the arrangement disintegrated into jealous bickering. Agostinelli left Proust and took up flying. In an unsuccessful attempt to win him back, Proust bought him a plane. Agostinelli enrolled in flight school under the pseudonym Marcel Swann—Proust had just published *Swann's Way*—and in May 1914 he crashed in the sea off Antibes near Monte Carlo when his wingtip caught the water on a low turn. Agostinelli did not know how to swim. He stood on the sinking wreckage a few hundred yards from shore and, in full view of the stunned crowd on the beach, drowned as a rescue boat approached. In *Remembrance of Things Past,* Proust modeled his narrator's great love, Albertine, a closeted lesbian who dies in a riding accident, on Agostinelli. The epic novel evoked France's affection for the new flying machines: "The airplanes which a few hours earlier I had seen, like insects, as brown dots upon the surface of the blue evening, now passed like luminous fire-ships through the darkness of the night. . . . And perhaps the greatest impression of beauty that these human shooting stars made us feel came simply from their forcing us to look at the sky, towards which normally we so seldom raise our eyes."

While Santos-Dumont could be gregarious in the company of male friends like Sem, he was diffident, to the point of being almost mute, around women. When he was a boy, he showed no such timidity: His main childhood playmates and confidantes were his sisters Virginia, seven years his senior, who taught him to read, and Sophia, two years his junior, his closest sibling in age, who died at ten of typhoid fever. But as a young man, although he was affectionate toward nieces and grandnieces, he was stoically silent in the presence of older female relatives. "He was quite odd," Amália Dumont, the wife of his eldest brother, told a reporter. "He'd sit in the most distant chair in a social circle. He'd fold his arms, lower his head, and remain that way for hours if necessary."

As Santos-Dumont was one of the few Parisians who owned an automobile, women who wanted to experience their first car ride were drawn to him. He usually obliged them, and on one occasion his chauffeuring unwittingly embroiled him in a domestic dispute. On January 16, 1903, the *Herald* ran a curious story under the headline "Santos-Dumont Named in Divorce Suit":

Boston Mass., Thursday—The name of Santos-Dumont was dragged into the divorce case of L. E. P. Smith, a life insurance company manager, against his third wife today.

The little thirteen-year-old son of Mr. Smith was recalled to the stand to testify how his stepmother made the acquaintance of "the balloon man" in a café in Paris and went automobiling with him. The boy said he identified the "balloon man" by seeing his picture in the paper.

The lad and his mother were seated in the café having supper. The "balloon man" was seated at a nearby table and entered into conversation with Mrs. Smith. Finally he invited her to go out in his automobile and she accepted. She sent her stepson home alone, according to the boy, and did not herself appear until ten o'clock the next day.

When Mrs. Smith went on the witness stand this afternoon she denied having been riding with Santos-Dumont or with any other man while in Paris.

Two days later, the *Herald* published Santos-Dumont's response: "There is not the slightest word of truth in the story . . . and I wish to deny it most emphatically. I have often driven members of the fair sex in my automobile, but never under circumstances which could justify any such story. . . . The whole thing is absolutely absurd."

As for Santos-Dumont's relationship with the beautiful young ladies who accompanied him to Paris social events, the man who thought so little of risking his life in front of crowds of thousands could not hold a simple conversation. The ladies complained to their friends that if he was not speaking about aeronautics, he did not have much to say. He was gentlemanly when helping them into a high chair at his aerial table or pouring them champagne at a public function, but he neglected to say good-bye at the end of the evening and left with no parting words or kisses. When he greeted them, he gave them flowers, but at the same time was too reserved to say hello, pronounce their names, or inquire after their health. Some of his female companions initially found his timidity to be endearing and convinced themselves that he did not have the confidence to make his romantic intentions known. But when the courtship did not progress, each woman was inevitably hurt and abandoned him or else tried to force his interest by whispering about impending nuptials to her friends and family. The "news" would spread quickly, and when it reached him, it never had the intended consequences: He simply did not speak to her again.

The tabloids occasionally ran short items about his betrothal to one society lady or another, along with his customary denial and angry pleading that the press not write about his private life. He gave the odd response that he would rather people think he was a widower than a fiancé. One story centered on Lillie "Lurline" Spreckels, the daughter of a San Diego sugar magnate, but only a few days after the papers reported the engagement, they announced that it had been called off because her parents found him too eccentric. There were similar retracted news stories in Paris about another American named Edna Powers. One of Santos-Dumont's friends from the 1890s, Agenor Barbosa, told a reporter decades later that although the aeronaut may not have had a love affair in his life, he was particularly fond of Miss Spreckels. The romance was probably never consummated, Barbosa suggested, because of his humiliation when her millionaire father publicly declared that aeronautics was too crazy and dangerous a profession for a married man.

On Santos-Dumont's visits home to Brazil, his intentions toward women were similarly puzzling. On the one hand, there was the recollection of his grandniece, Sophia Helena Dodsworth Wanderley: "He had no special lady friends. Whenever he came to Brazil, the newspapers announced his arrival—he was a celebrity. Men, women, and children surrounded him and pampered him. He was very courtly and well dressed, and very popular with the ladies. But he was never romantically involved. He was never even alone with a woman, only as part of a group of friends." On the other hand, there were the claims of Assis Chateaubriand, the "Citizen Kane" of Brazil—newspaper tycoon, ambassador to England, notorious Don Juan—who had boasted of seducing every woman he ever desired with the exception of a young comedic actress who resisted him and his proposal of marriage. He was heartbroken, he said, until he had a conversation with Santos-Dumont in which the aeronaut revealed that he too had been rebuffed by the enchanting comedienne when he asked for her hand in marriage. It was her problem, Chateaubriand proclaimed, if she rejected the conqueror of the air and the conqueror of the press. Years later the actress in question said that her relationship with Santos-Dumont was merely platonic. She said that she appreciated the orchids that he had lavished on her but wished that he had been a more talkative companion.

ON SEPTEMBER 7, 1903, Santos-Dumont returned to Brazil. It was Independence Day, and he received a hero's welcome, banquets and fetes that rivaled those in Paris, London, and New York. He could not walk in the street without strangers accosting him. But people could not understand why he did not fly in Brazil. Why had he not brought an airship with him from Paris? His countrymen did not accept his explanation that even if he had a balloon with him, an ascension was not something that could be accomplished on the spur of the moment. He needed a hangar and a hydrogen-generating facility. That raised the question of why he had established his workshop in Paris instead of in São Paulo or Rio. According to his nephew, Henrique Villares Dumont, he "was deeply distressed when the tongue of malicious gossip said that he was not only frenchified in his manners and mode of life but had actually become a French citizen." In fact he remained a Brazilian national and had gone out of his way in Paris to remind people of that by flying the Brazilian flag from his airships. He thought of himself as half French, half Portuguese, and started signing his name with an equal sign, Santos=Dumont, to indicate that neither heritage was more important. He was disappointed by his reception in Brazil, and, after only sixteen days, returned to Europe.

That fall he took a break from flying in order to write down his experiences thus far. He stayed mostly at his desk, except for the occasional social function and regular dinner breaks at Maxim's or at the high table in his apartment. The book, *My Air-Ships,* was strictly an aeronautical memoir, long on colorful descriptions of each of his flights and short on information about the rest of his life; there was no discussion, say, of who his friends were or what he did when he was on the ground. It was unusual for a thirty-year-old to write an autobiographical work, but his need to be understood, particularly by his own countrymen, compelled him to justify himself in print. Although half his life was ahead of him, he felt that he had already achieved his principal goal of a personal flying machine. He had gone everywhere in Paris in *No. 9,* and even if the airship only worked in calm weather, he thought that his job was completed. He had shown how it could be done. That was enough. Others could adopt the next generation of internal combustion engines so that the airship would no longer be defeated by the wind. In 1903, he was the aeronautical darling of Paris and the unrivaled master of the sky, but there was competition on the horizon. He was oblivious to it but so was the rest of the world.

Far across the Atlantic, on the shores of Kill Devil Hills, four miles south of Kitty Hawk, North Carolina, Orville and Wilbur Wright were taking turns making quick hops across the sand in the world's first airplane, with Wilbur going the farthest, 852 feet in fifty-nine seconds, on December 17, 1903. To evade potential competitors, the Wright brothers had worked in near secrecy, choosing Kill Devil Hills for both its favorable winds and remoteness. They were determined to be not only the first to achieve powered flight but also the first to build a plane good enough to sell to one of the world's militaries. The secrecy worked. A few men who worked as "lifesavers" at Kitty Hawk—who were on the lookout for shipwrecks in the days before the Coast Guard—did see the flights but did not initially speak to the press. The Wrights' historic first flights, and subsequent ones in Ohio over the next two years, received little publicity. Indeed, the first journalist to watch them pilot the Flyer biplane in Ohio wrote up what he saw in a magazine for apiarists, *Gleanings in Bee Culture,* and the account did not appear until more than two years after Kitty Hawk. No other invention of monumental importance was ushered into the world so quietly. Even the clandestine bomb-building at Los Alamos received more notice.

Today we take it for granted that the Wright brothers invented the airplane, but the situation was not clear in the early 1900s. When Santos-Dumont finally fulfilled his promise to Samuel Langley to experiment with heavier-than-air flight and piloted a biplane in 1906, three years after Kitty Hawk, he was hailed in France and elsewhere in Europe as the inventor of the airplane. The Wrights brothers' stealth and the absence of official witnesses were only part of the explanation of why their work had received so little attention. Equally important was the tendency of the American press to discount claims of aeronautical prowess because they had so often proven to be spurious. The biggest blowhard, the press concluded, was Langley himself.

On December 8, 1903, six days before the Wright brothers tested their plane for the first time, the sixty-nine-year-old Smithsonian director led co-workers and witnesses from the army to a wide and somewhat secluded section of the Potomac to view the Great Aerodrome. The 720-pound plane with two sets of wings, each spread upward in the form of a squashed letter V, was mounted atop the catapult on a houseboat moored in the middle of the river. Charles Manly, a mechanical genius who was Langley's assistant, had redesigned the Balzer petroleum engine into a 52.4-horsepower monster that could run for ten hours—a major achievement in an age when plodding car engines boiled over after only an hour or two on the road. Manly had enough confidence in his work that he planned to pilot the Great Aerodrome himself.

He had actually tried two months earlier, on October 7. It was supposed to be the Aerodrome's maiden flight, but the catapult failed and the heavy plane, with the engine going full throttle, barreled down the launch rail and dove into the water. Manly managed to slip out of the submerged cockpit and swim to shore, and the aircraft was pulled from the river. Langley inspected it and declared that there was no problem whatsoever with the plane. The problem, he said, was confined to the catapult. Nothing a simple redesign would not take care of. He and Manly vowed that the Aerodrome would fly.

The press was not as hopeful. Newsmen had little sympathy for Langley because of the secrecy that had surrounded his previous work. The *Washington Post* reported that the plane had flown as well as "a handful of mortar." The *Boston Herald* suggested that the Smithsonian director should build submarines instead because his machines had a natural affinity for water. Washington writer Ambrose Bierce mocked Langley's placing the blame on the catapult:

An Ingenious Man who had built a flying-machine invited a great concourse of people to see it go up. At the appointed moment, everything being ready, he boarded the car and turned on the power. The machine immediately broke through the massive substructure on which it was built, and sank out of sight into the earth, the aeronaut springing out barely in time to save himself. "Well," said he. "I have done enough to demonstrate the correctness of my details. The defects," he added with a look at the ruined brickwork, "are merely basic and fundamental." On this assurance the people came forward with subscriptions to build a second machine.

Wilbur Wright followed the press accounts. "I see that Langley has had his fling, and failed," he wrote to a friend. "It seems to be our turn to throw now." In another letter, written on the eve of Langley's December trial, Wilbur said, "It is now too late for Langley to begin over again."

Langley did not know then that the secretive Wrights were a few days away from success at Kitty Hawk. He knew of their reputation as glider pilots but, like most of the aeronautical community in the United States and abroad, he was not aware of how close they were to powered flight. Although he was not knowingly in a race that cold December afternoon, he was eager to redeem himself after October's fiasco.

Manly was in the cockpit. He was impatient to take off because the wind was picking up, darkness was approaching, and the air was uncomfortably cold. At 4:45 P.M., Langley gave the signal and the Great Aerodrome sped down the catapult's sixty-foot track. The ungainly plane climbed a few feet into the air, flipped upside-down, and plunged nose-first into the Potomac. Manly feared that he was going to drown and struggled to extricate himself from the submerged wreckage. His clothes had frozen to his body. A doctor in attendance had to cut them off. When Manly was revived, he let out a series of blasphemies in earshot of his crestfallen and expletive-sensitive boss and the group of distinguished witnesses.

The unsuccessful flight received more press than the previous trial. "Langley's Dream Develops the Qualities of a Duck," ran the headline in the *Raleigh News and Observer*. "It Breaks Completely in Two, but Without Even An Expiring Quack, Drops a Wreck into the Icy Potomac," the subhead continued. "Perhaps if Professor Langley had only thought to launch his airship bottom up, it would have gone into the air instead of down into the water," another paper said. "The professor does not have sufficient faith in his work to risk his life in the machine when the attempts to fly it are made," protested the *Wilmington Messenger*. "He either goes to Washington City or places himself at some safe distance when the attempts are made."

But it was the coverage in the country's most prestigious papers that upset Langley the most. The *New York Times* and the *Washington Post* both called on Congress to cut its losses. "In the past," said the *Post*, "we have paid our respects to the humorous aspects of the Langley flying machine, its repeated and disastrous failures, the absurd atmosphere of secrecy in which it was enveloped, and the imposing and expensive pageantry that attended its various manifestations. It now seems to us, however, that the time is ripe for a really serious appraisement of the so-called aeroplane and for a withdrawal by the government from all further participation in its financial and scientific calamities."

Langley continued to insist that the problem was not the plane but the launching device. "Failure in the Aerodrome itself or its engines, there has been none," he said, "and it is believed that it is at the moment of success." Congress was not convinced. "Tell Langley," said Representative Joseph Robinson, "that the only thing he ever made fly was government money." Representative Gilbert Hitchcock's denunciation was equally personal: "If it is going to cost us $73,000 to construct a mud duck that will not fly 50 feet, how much is it going to cost us to construct a real flying machine? I realize that Professor Langley is a learned man. He knows a vast amount about extinct animals and stuffed birds. But I see no reason why at national expense he should be constituted a modern Darius Green who met with no more complete failure in his attempted flight from the New England barn than this modern imitator at national expense has attained." Langley soon withdrew from the public eye, his health declined rapidly, and he died a defeated man in 1906.

But Langley had done more for aeronautics than he realized. In 1899, the young Orville and Wilbur Wright, not knowing where to read about the history of flight, wrote to him in his capacity as director of the Smithsonian and asked for a reading list. The Wright brothers later credited the Smithsonian's response as inspiring them to build a flying machine. "The newpapers report the death of Professor Langley," Wilbur wrote to Chanute. "No doubt disappointment shortened his life. It is really pathetic that he should have missed the honor he cared for above all others. . . . The fact that the great scientist, Professor Langley, believed in flying machines was the one thing that encouraged us to begin our studies." Little did Wilbur know when he wrote these gracious remarks that his brother was about to become embroiled in a messy, thirty-year dispute with the Smithsonian.

Charles Walcott, an eminent paleontologist, who succeeded Langley as head of the Smithsonian, was eager to house the Flyer in the museum but would not agree to label it as the first plane *capable* of carrying a man. He believed that the Aerodrome deserved that description, accepting his predecessor's view that the craft was airworthy and would have flown were it not for the faulty launching mechanism. In 1914, Walcott irked Orville (Wilbur had died two years earlier of typhoid) by lending the remains of the Aerodrome to Glenn Curtiss, the Wrights' unscrupulous rival who owed royalties for infringing on their patent. Curtiss hoped to avoid payment by demonstrating that Langley was the true inventor of the airplane. Using funds put up by the Smithsonian, Curtiss rebuilt the Aerodrome but quietly changed the design, strengthening its wings and adding floats. On May 28, 1914, Curtiss flew Langley's aircraft for less than a minute over Lake Keuka near Hammondsport, New York. The judge overseeing the patent dispute took little notice but Langley's followers seized on the stunt. The museum put the Aerodrome back on display, labeling it as "the first man-carrying aeroplane in the history of the world capable of sustained free flight."

Disgusted by the Smithsonian's refusal to recognize the Flyer's priority, Orville exhibited the plane at other American institutions before shipping it off in 1925 for permanent display at the London Science Museum, which promised to properly credit it. "I believe that my course in sending our Kitty Hawk machine to a foreign museum," said Orville, "is the only way of correcting the history of the flying-machine, which by false and misleading statements has been perverted by the Smithsonian Institution. . . . With this machine in any American museum the national pride would be satisfied; nothing further would be done and the Smithsonian would continue its propaganda. In a foreign museum this machine will be a constant reminder of the reason of its being there. . . ."

It was not until Walcott died in 1927 that the Smithsonian, under its new director, Charles Abbott, tried to make amends with Wilbur, but the overtures he made did not go far enough in acknowledging the institution's responsibility in distorting the history of flight. Fifteen years later, in the middle of the Second World War, Abbott finally offered an acceptable apology—just in time, because Orville was ill and the Smithsonian may never have received the plane if he had died before amending his will. In 1948, when the war was over and Orville had been deceased a year, the Flyer was moved from London to Washington.

The dispute over whether the Aerodrome was airworthy was not settled for three more decades. In 1982, the Smithsonian enlisted the help of NASA to determine whether the Aerodrome could have flown without the "improvements" Curtiss had introduced. Engineers put the plane through stress tests and concluded that it was structurally too weak and would have quickly broken apart in the air. They identified more than eight places where it would have snapped. "The two metal tubes supporting the fuselage, for instance, couldn't take any twisting force," said Smithsonian engineer Howard Wolko. "And the beam bearing the load of the wings is circular, which is about the weakest shape that you can use. It's the strangest structure that I've ever seen." Perhaps it was just as well that the museum did not reach this conclusion during Langley's lifetime. He had suffered enough ridicule and died with the comforting if mistaken impression that his aircraft could fly.

The critical press that the Aerodrome received in early December 1903 did much more than injure Langley's pride. The mocking stories in the world's leading newspapers turned public opinion against the idea that *any* plane could fly. Indeed, when word leaked out a week after Langley's humiliation that the Wrights had flown at Kitty Hawk, the press was skeptical. Few people were prepared to believe that two small-town bicycle salesmen could succeed where the dean of American science, with the backing of Congress and the Smithsonian, had failed so miserably. After all, Langley had spent twenty thousand dollars on his defective catapult whereas the Wrights had spent a mere four dollars on their launch rail. It did not help that the Wright brothers had not invited any newsmen along or granted any interviews.

The *Virginia-Pilot*, the first paper to report the events at Kitty Hawk, was tipped off by an intercepted telegraph that the brothers had sent home to Dayton from the Outer Banks. The story ran on December 18, under the banner front-page billing: "Flying Machine Soars Three Miles in Teeth of High Wind over Sand Hills and Waves at Kitty Hawk on Carolina Coast." Below the banner read, "No Balloon Attached to Aid It. Three Years of Hard, Secret Work by Two Ohio Brothers Crowned with Success. Accomplished What Langley Failed At. With Man as Passenger, Huge Machine Flew Like Bird Under Perfect Control. Box Kite Principle With Two Propellers." The hyped-up story (it had flown less than a fifth of a mile) went out over the Associated Press wire. The Wrights found the coverage in their hometown paper particularly galling. The *Dayton Daily News* mistook the Flyer for a dirigible and buried the AP story in the section devoted to local news under the headline "Dayton Boys Emulate Great Santos-Dumont."

[C H A P T E R   1 2]

A   S C U R R I L O U S   S T A B B I N G

A N D   A   R U S S I A N   B R I B E

S T .   L O U I S ,   1 9 0 4

•

I N 1 9 0 4 , most of the mainstream press in America was not convinced that flying machines would ever be a commercial success. On the eve of the aerial competition at the St. Louis Exposition, the *New York Times* branded the contestants as showmen and declared that "the chief interest which the tests of flying machines will have for visitors to the fair grows out of the probability that a majority of the machines tested will go to pieces in one way or another and spill their conductors. Even among those who would regret such an accident, a great many would probably confess that if it must happen they would like to see it." But those who insisted on witnessing crashes, the *Times* continued, need not crane their necks toward the sky. "The automobile is sufficiently dangerous to satisfy all reasonable ambitions in the matter of foolhardiness. When it has been made quite safe our contingent of persons afflicted with motor madness may take to ballooning; but that is still a long way off. The automobile offers so many and various opportunities for self-destruction that until its possibilities have been exhausted there will be no eager demand for anything else in the way of a mechanical fool-killer."

St. Louis was a good location for an aeronautical competition. Not only was its terrain agreeable (no mountains to smash into, no large bodies of water to drown in), its heritage was appropriate too. The city had been home to John Wise, the greatest American balloonist of the nineteenth century. Born in 1808, six decades before Santos-Dumont, Wise too got hooked on aeronautics as a child. He sent kittens aloft on kites, and tested a tissue-paper parachute by dropping a cat out of the window. The animals survived, but his neighbors would not let him near their pets. His experiments with *montgolfières* were less successful. When he was fourteen, he released a paper balloon with a fire in its basket; after ascending a few hundred feet, it came down and ignited a neighbor's thatched roof. Mishaps like this forced him to give up aeronautics until his late twenties, where after dabbling in theology, cabinetry, and piano construction, he decided to become a professional aeronaut and build his own balloon, even though, as he later confessed, "I had never seen an ascension with one, nor had I any practical knowledge of its construction. . . . I intended merely to satisfy my desire of sailing aloft to enjoy a prospect that I had ever considered must be grand and sublime."

Wise's most important contribution to the science of ballooning was the "rip panel," a seam in the balloon that could be opened at the pull of a cord if gas needed to be vented in an emergency. The rip cord was also routinely pulled on landing so that the balloon would deflate before the wind tossed it about on the ground. Wise's invention was serendipitous. On August 11, 1838, his balloon had burst at thirteen thousand feet under pressure of the expanded hydrogen. Rather than plummeting to his death, he was amazed to find himself descending slowly as the ripped envelope bunched up at the top of the rigging and acted as a makeshift parachute. It took him two months to duplicate this accident in a controlled way with his new rip panel.

Wise was a sportsman, like Santos-Dumont, with an eye on the record books. In July 1859, he made his 240th ascension in the *Atlantic,* the largest balloon in the world. He and three companions set out from St. Louis hoping the wind would carry them all the way to New York or another East Coast city. The Brazilian would have approved of their attention to culinary provisions: "a large quantity of cold chickens, tongue, potted meat, sandwiches, etc.; numerous dark colored, long-necked vessels, containing champagne, sherry, sparkling ca-tawba, claret, madeira, brandy and porter." The alcohol may have made the ride seem smoother but it may also have been responsible for a mistake that almost cost Wise his life. At midnight on the first day, he bid his companions goodnight and lay down to rest with his head directly under the neck of the balloon. The *Atlantic,* with a gas capacity of 120,000 cubic feet, had been only half filled with hydrogen on the expectation that the gas would expand at high altitudes. Sure enough, while Wise was sleeping, the balloon steadily gained elevation and the hydrogen inflated the balloon to the point where it was on the verge of bursting. Gas was now rushing out of the neck of the balloon into Wise's face. His noisy, erratic breathing alerted his colleagues, who pulled him away, shook him awake, and saved him from asphyxiation. This was one kind of freak accident that Santos-Dumont, the king of aerial mishaps, never experienced.

The rest of the trip was eventful, too. A hundred-mile-an-hour gale swept Wise and his friends off course for fifty miles over Lake Ontario. When they finally regained control of the balloon and tried to land in a forest on the shore, they were still traveling too fast for the anchor they threw out to grip anything. It bounced off one tree and broke its tines on another. Then the balloon ripped, and after a mile of scraping the forest canopy, the *Atlantic* and its four bruised passengers came to rest in treetops twenty feet above the ground. Wise pulled out his pocket watch and proudly announced that they had been traveling nineteen hours and forty minutes. Although they had made it only as far as Henderson, New York, their 826-mile journey set a distance record that would not be broken for forty-one years. (In October 1900, Count Henry de la Vaulx flew 1,193 miles from Paris to Korosticheff, Russia, in thirty-five hours and forty-five minutes.)

Wise made his 463rd ascension at the age of seventy-one. On September 28, 1879, he set off from St. Louis in a small and flimsy-looking balloon called *Pathfinder*. He was accompanied by a young bank teller named George Burr. Before they departed, Wise questioned *Pathfinder*'s airworthiness and reportedly warned Burr: "If only one man can go, I will be that man; if two can go, you will be the other one; but I would rather leave you behind. I am old enough now to die: you are young enough to live many years." His words were prophetic—the men never returned. A month later a decomposed male body washed up on the southern shore of Lake Michigan; the shirtsleeves were monogrammed G.B. Wise's body was never recovered.

St. Louis's small aeronautical community saw the 1904 competition as a fitting tribute to the man who a quarter of a century earlier had been the first champion of aeronautical science in the United States. In January 1904, the fair's organizers settled on the rules for the summer's $150,000 competition, and Santos-Dumont gave them his blessing. In the marquee event, a winner-take-all prize of $100,000 would go to the flying machine of any description, manned or unpiloted, lighter or heavier than air, that achieved the best total time for three trials on a ten-mile triangular course provided that the average speed on each trial exceeded twenty miles an hour. The trials had to be completed by September 30, and the only prerequisite for entering a machine was that it had to have already made a round-trip flight of a mile or more. Santos-Dumont announced his intention to compete and predicted that he would have substantial company: "I expect that at least 150 airships will be entered when the rules and conditions of the contest are known."

Not everyone shared his sanguine opinion. In March, Leo Stevens, an aeronaut in New York, informed *Scientific American* that he was not going to participate. "The speed expected is too great," Stevens wrote. "The man who enters has everything to lose and nothing to gain. The rules call for a speed of at least 20 mph. This is impossible. The prize is perfectly safe with the Exposition Company. I think the rules might have been modified just a little. For instance, the man making best time should be allowed to take first prize, second man second prize, third man third prize. There would then be something in sight."

Santos-Dumont spent the first three months of 1904 in New York in a suite at the Waldorf-Astoria. From the chandelier in the living room, he hung a fifteen-foot, one-tenth-scale replica of the airship he planned to take to St. Louis. The housekeeping staff, a head taller than Santos-Dumont, had to duck when they cleaned the room.

A visitor who admired the model asked Santos-Dumont whether he derived much pleasure from his flights. "More than the most ardent automobilist finds in his favorite racer," he replied. "Of course there is a pleasurable sensation in going through the air at full speed. But that is not it. The great sensation is being able to command a machine fifty mètres long while sweeping through the air. That is indescribably delightful." He laughed as he drew himself up and displayed his slender physique. "I'm not very big or very powerful," he continued, "but when I'm standing in my basket, that machine has to obey me. I am not controlled by it, but I command it. It's the realization of that sense of power that makes air navigation a fascinating pursuit."

In interviews in New York, he questioned whether the St. Louis fair really had the $100,000 to award and suggested that if the money was not put in escrow, he would reconsider his plans. The organizers were counting on his presence to add legitimacy to the competition, and they were incensed by his questioning their finances in public and by his failure to come to St. Louis to discuss the matter in person. The exposition would be the largest world's fair to date, backed by $50 million from Washington and private industry, so there should not have been any question about the validity of the prize. But Santos-Dumont did not stop there. He sent a cablegram to the organizers insisting that they pay him $20,000 to participate. This, he said, was the cost of building *No. 7* and of shipping it and three mechanics to St. Louis. He demanded that the payment be made in secret. Now the organizers were in a quandary. They needed the participation of the only aeronaut in the world who was a household name, but they were not going to treat one contestant favorably—and certainly not one who spoke unkindly about them to the press. They also feared that the unstable Brazilian might make public any supposedly private arrangement.

Santos-Dumont backed down when the fair sent an emissary to the Waldorf, where the aeronaut had secluded himself to read French poetry and order rich Parisian cuisine from room service. He even instructed the Waldorf kitchen how to make a leek sauce in the same compulsive way he would direct his workmen. He told the emissary that he was enjoying New York. "They even accuse me of growing stout," he said. He asked the man to relay one request to the fair's organizers: reduce the required speed from twenty miles per hour "to a trifle under 19." Despite all his bravado about having the best airship, he knew from his limited tests in Paris that he was not assured of winning.

The organizers cheerfully modified the rules. They did not have much choice if they wanted a serious competition. Although more than eighty people were signed up for the various aerial events, only eight had paid the $250 fee that enabled them to vie for the grand prize. Two of the eight soon dropped out because they had misunderstood the rules. And of the remaining six, close scrutiny by the judges would reveal that none besides Santos-Dumont had met the one-mile prerequisite.

On March 22, Santos-Dumont left the Waldorf without giving the hotel advance notice and returned to France to check on the construction of *No. 7*. Evidently he did not like what he saw because on May 16 he cabled the fair's president: "Lost sixty horse power engine. Only able get forty. Airship tried yesterday. Only goes less than twenty miles. Cannot race until speed condition is cut down to fifteen mph." At the same time, the Paris Aéro Club, eager that Santos-Dumont not walk away with $100,000, was lobbying St. Louis not to offer one grand prize but a series of much smaller prizes.

The organizers conferred. So far Santos-Dumont was the only experienced flier in the competition, and they were eager for it to take place if they could relax the speed requirement again without humiliating themselves. As for the Paris Aéro Club, the organizers did not believe that smaller prizes would draw more competitors all the way from Europe. After spirited debate, they announced that the prize of $100,000 would remain for a speed of twenty miles an hour but that a reduced prize of $75,000 would be awarded if the winning machine achieved only eighteen miles an hour and $50,000 for fifteen miles an hour. Santos-Dumont said he was satisfied.

On June 12, he set sail from Le Havre for New York aboard the steamer *Savoie*. With him were his mechanic Chapin and two assistants, Gerome and André. They were accompanied by three large wooden crates that weighed four thousand pounds and contained the disassembled pieces of *No. 7*. The party boarded a train in New York and reached St. Louis during the last week in June. "I have never raced this airship," Santos-Dumont told the *New York Times,* "and have had but three trials with her in Paris. They were for short distances, but everything worked admirably. This machine is much stronger and much more powerful than the *No. 6* in which I circled the Eiffel Tower, and, though I have never timed her, I feel sure that she will fulfill the requirements." Although which requirements he had in mind were not clear because at the same time that he made these remarks he was pressing yet again to change the rules. He had inspected the triangular course, whose shape he had suggested, and, to the consternation of the organizers, said that he would now prefer to fly in a straight line. The triangular course had two time-consuming turns; a straight round-trip course would require only one turn, when the aeronaut brought his ship around to steer it back to the starting point. Once again the fair accommodated his wishes, and he announced that he would make a bid for the prize on the Fourth of July.

On June 27, a legion of customs inspectors watched Santos-Dumont and his workmen unpack *No. 7*. When the officials assured themselves that there was no contraband hidden in the folds of the silk, Santos-Dumont examined the pieces of the airship and proclaimed that everything was in good order. In Paris, the balloon envelope, consisting of two layers of silk glued together, had been varnished seven times, two coats on the inside and five on the outside, to make it impermeable to air. For good measure he planned to apply an eighth coat in St. Louis, but first he wanted to let the oily silk air out. He left the balloon overnight in its crate with the top off. Mindful of the vandalism at the Crystal Palace, the exposition had hired a security force from a local military base, the Jefferson Guards, to patrol the balloon shed and other exhibition buildings. A man named J. H. Peterson stood guard until midnight, when he was relieved by Lucian Gilliam, who was still on duty when Santos-Dumont's crew returned at 7:00 A.M. One of the workmen noticed that the balloon had been slashed in four places. The cuts were two to three feet long, and because the balloon had been bunched up, the cuts went through not only the visible silk but also the layers underneath. There were forty-eight gashes in total.

Carl Meyers, the aeronautics expert in charge of the contest, was the first exposition official to arrive at the shed. "To me the cutting was done with a dull jack knife," he said, "with no other object but malicious destruction. I can repair the damage, but it will probably take two weeks, and possibly longer. I cannot tell definitely until the bag is taken from the crate and spread out."

When Santos-Dumont was awakened with the bad news at the Hamilton Hotel, he broke into tears. "It is an outrage! An outrage!" he cried. "I cannot imagine who would have done such a thing. I have no enemies over here. It must have been the work of some crank."

The recriminations started at once. The fair officials reminded him that they had warned him not to leave the top off the crate. Santos-Dumont in turn accused them of not adequately watching the balloon because Gilliam had admitted that he had left his post twice, at 2:00 A.M. and 4:00 A.M., to get a cup of coffee at the guard headquarters, some three hundred feet from the aerodrome.

The watchmen remembered seeing a "nervous man" who had visited the shed many times over the course of a few days. The St. Louis police tracked him down and detained him, but he turned out to be a harmless nut who had visions of escaping his earthly troubles by permanently launching himself into space. Santos-Dumont joked that he could appreciate the man's fantasies. The man explained that he had loitered around the shed because he wanted to meet the famous Brazilian and persuade him to take him aloft. He possessed a small jackknife, but this by itself was not incriminating because many people carried them in 1904. The man did not seem to be the violent sort, and he had an alibi. When the forensic examination turned up no traces of varnish or silk on the blade, the police released him, and the exposition offered a one-thousand-dollar reward for information leading to the arrest of the real culprit.

The fair officials were eager to have the balloon repaired locally, so that the race would not be delayed by more than a week or two. But Santos-Dumont had other ideas. "The bag of the airship is made up of numerous squares, each one sewed in place, glued, varnished and specially prepared for the purpose," he explained. "It will take three or four men and women several weeks to make the necessary repairs, and I will be content with none but French workmanship," which would cost him between five and eight thousand dollars. The consensus of the other aeronauts at the fair was that he was "so wrought up that he overestimates the damage." To break the stalemate, Meyers offered to repair the balloon at his own expense, a proposal that Santos-Dumont dismissed out of hand. "If Professor Meyers repairs the bag, he can go up in the ship himself," he said. "I will not trust myself in it."

The standoff worsened when Colonel Kingsbury, chief of the Jefferson Guards, issued his official report. After describing in detail the movements of his men on the night in question, and explaining that he had fired Gilliam for briefly abandoning his station, Kingsbury accused Santos-Dumont of contributing to the security lapses. He said it was common knowledge that the fair had the resources to assign only one night guard to the expansive balloon shed and that Santos-Dumont had ignored suggestions that he supplement the security by stationing one of his own men next to *No.* 7. "I learn from Lieut. Walsh of the Secret Service, who has investigated the matter thoroughly, that Mr. Hudson, Superintendent of the marine section of the transportation bureau, told him that he several times yesterday spoke to Mr. Dumont of the advisability of placing special watchmen over the balloon and placing the cover on the box," Kingsbury reported. "This Mr. Dumont failed to do, only partially replacing the cover, as he stated he desired to get all the air possible in contact with the balloon. If air had been essential to the balloon, as much could have been obtained if a strong wire netting had been placed over the top of the box, the cover of which was only half removed anyway." And then Kingsbury made a damning observation: "One of Mr. Dumont's assistants is said to carry a large, strong knife. Owing to the strength of the numerous folds and thickness of the balloon envelope, some such knife as this would have been necessary to cut the balloon."

The implication was shocking: Santos-Dumont had arranged for the destruction of his own balloon. For the first time since he had won the Deutsch prize, he made front-page news on both sides of the Atlantic. And the press uncovered the fact that his balloons had been slashed not just in St. Louis and London but once in Paris. Was it really possible that a deranged vandal had followed Santos-Dumont around the world? Or was it more likely that the aeronaut himself was somehow responsible?

Santos-Dumont departed for France hours later and wrote a long, indignant denial to American papers:

How is it conceivable that I destroyed my balloon, which is my bride, my adored, my idol? I have devoted my life to the conquest of the air. That I have risked my life I need not tell you, my friend, who saw me fall to the roof of the Trocadéro, who saw me plunged into the Mediterranean and who despaired of my safety a hundred times in Paris.

Here, then, is the proposition. I have expended my income, have gambled with my existence, have tried and failed, and tried again and succeeded in some measure, and I still struggle on. I won the great Deutsch prize in Paris, given to the one who should fly around the Eiffel Tower within a time limit. I gave the proceeds to charity.

In the new world they offer as the crowning feature of the greatest Exposition ever held a vast reward for the aeronaut who shall cover a given course within a given time. I construct with great pains and cost an airship and bring it to St. Louis, together with my three mechanics. I, who have compelled recognition in the old world, am anxious to win the praise of the new.

Who does not desire the plaudits of a people so fertile in invention, so like the ocean in energy, so quick to welcome the new? All that I have done in Europe is a far cry. I must demonstrate to Americans what I have accomplished in the air. As you say in your expressive language, I must "show."

I arrive here. I begin my preparations for an ascension. The silk bag of my balloon is destroyed in the night by some miserable malefactor of whose identity I have no conception. Is it conceivable that I should do this thing? The charge is supremely ridiculous. Yet I must confess it is annoying, because it argues such stupidity on my part.

Really, my friend, it is such discouragements as this that try the inventor, the pioneer, much more than his failures. In Europe, in South America, they laugh at this story. But there may be some in America who may harken, and this grieves me. Why should I shirk an ascension in St. Louis? Is it because I am afraid? I have made three thousand ascensions and have met with every disaster, except death, that befalls the air sailor, and I am as ready to ascend today as I was when I won the Deutsch prize.

Is it because I fear failure? Who are my antagonists that I shall be alarmed? I have failed before, failed many times. That is the lot of the experimenter. Failure in aeronautics is in no sense a disgrace. No hard and fast rules are laid down. We have no knowledge of conditions.

One knows how fast a horse must run a mile or a sprinter the hundred yards to be great, but who can settle metes and bounds for the balloonist? He does the best he can and trusts to the good God.

No, there is no earthly reason why I should wish to evade an ascension in St. Louis. The prize of success is princely and ample time is given for trials. But forgetting the prize, which if I won, I would give to charity, consider the glory of the triumph. It would be recorded through all the ages in which the story of aeronautics was told, and perhaps the last intrepid navigator in the aerial ocean would speak the name of the winner at St. Louis—the Columbus of the atmosphere.

Frankly, I love glory. I desire fame. Is it credible then that I should reject the supreme opportunity of my life? I have been consulted by the commanders of armies regarding the use of the balloon in warfare. I was asked by Japan to join her forces in Corea as head of the balloon service, and a fabulous offer was made me to take this very Santos-Dumont *No.* 7 to the front and essay to drop high explosives into Port Arthur.

I was sorely tempted, but many of my best friends are Russians, and much as I admire the Japanese, I am compelled by the ties of Caucasian kinship to deny my aid to the yellow man. Perhaps I could have done nothing. That is known only to the war god.

France has adopted my plans for military balloons and in her next war will use them. Your World's Fair president, Mr. Francis is a gentleman. He has the savoir faire, the charm of a great diplomat. He resents the charge made by his employe and has assured me in a private letter of his distinguished consideration.

All this sounds very egotistical and is distasteful for me to say, but it is necessary for me to remind you who it is that a Jefferson guard, perhaps asleep at his post, charges with being a sneak, a coward and all that the wretched act at the Aerodrome implies.

It is even insinuated that my men, my beloved Chapin, Gerome, and André, who have toiled with me for four years, did this thing. They are furious to-day over the insult, but they are better gentlemen than the uniformed employe who has dared to asperse them.

Their earnings are more than that of a whole company of the Jefferson guards. When I won the Deutsch prize, they each received $1,000, and it was understood by them that if I was victorious in St. Louis they were to receive $5,000 apiece. When I win they win.

It is charged again that I wanted a concession. It is the nadir of imbecility. I am not in trade. There are provinces in São Paulo growing coffee for me. I am not a scorner of trade. That is weakness, but, nevertheless, I am not seeking the public's money.

As a matter of fact, Mr. Skiff [the director of exhibits] came to me and asked that I consent to allow the balloon to be shown for an admission fee. He said that the Exposition needed the money and that it would be a good thing for the Exposition and me. This I refused to do.

I am not the balloonist at a country fair. I am here on a great enterprise, and I cannot divert my energies in any side line whatever. It is your Fair which is seeking concessions, not I. This is one of the times when one realizes that there are problems that the code duello alone may solve.

Santos-Dumont's wordy response was a mixture of justification, exaggeration, and wishful thinking, and it exposed his conflicting sentiments about prize money. On the one hand, he had campaigned for a substantial award on the ground that his experiments were expensive. On the other hand, he was quick to let people know that he really did not need the money. That he wanted it only for his already well-paid workers. For Santos-Dumont the size of the prize was a measure of his self-worth. He was focused on his place in history. If the leading aeronautical clubs and world fairs grandly remunerated his record-setting efforts, it confirmed his belief that he was doing something important. And yet there was another contradiction in his attitude toward money. Although he said he did not want to be a circus performer who charged admission for his flights, he had struck a secret deal with the Crystal Palace exhibitors in which they sold tickets to view his slashed balloon and passed on to him a healthy share of the receipts. The stabbed gasbag, during the brief time it was on display in London, with tasteful bandages covering its wounds, proved to be a popular attraction; people queued up for hours to see it. Despite his comment about turning down Skiff's request to exhibit *No. 7,* he did in fact try to make a similar, under-the-table arrangement in St. Louis, but high-minded exposition officials rejected it as an unseemly way to profit from the vandalism.

As he often did when he was under pressure, Santos-Dumont twisted facts in his denial: He had long stopped receiving money from coffee, and the three thousand ascensions he claimed were overstated. And although he wanted to believe that speculation about his involvement in his balloon's destruction was confined to America, it was the talk of Europe too. His tendency to distort the truth means that the historian must not uncritically accept everything that he said in his defense. His statement about Japanese interest in his airships, for instance, strains credibility because he had never mentioned this remarkable fact before, albeit there had long been rumors about foreign militaries approaching him. One could completely dismiss the statement if it did not figure in an extraordinary charge made by the Jefferson Guards.

Although the guards initially suggested that Santos-Dumont had slashed his own balloon to save face from failure, later they unofficially offered a more complex motive. The Japanese government, they claimed, had promised him $1 million if, after demonstrating *No. 7*'s mettle by winning the prize in St. Louis, he gave *No. 7* and two other airships to the Mikado's army for use against the Russians. An agent from Moscow, they said, offered Santos-Dumont $200,000 to break the contract with the Japanese. Not confident that he could pull off the flights necessary to earn the million, he "jumped at the Russian offer of $200,000 in cold cash and cut his balloon to pieces." Santos-Dumont said it was beneath his dignity to respond to such a scurrilous accusation, and the governments involved never had any comment either.

He could have silenced his critics by following through on the idea of repairing the balloon in Paris and returning with it to St. Louis. But he told a few friends of a far more ambitious plan to salvage his reputation and ensure himself a larger place in history: he would build the world's first airplane.

In the end, no one won the grand prize in St. Louis. Of the $150,000 earmarked for aerial contests, the exposition parted with only a thousand dollars, in a kite competition and a staged balloon exhibition. The lack of winners proved that the contest rules were too rigid and that Santos-Dumont, even though he did not fly in Missouri, was in a league of his own. His airship, to be sure, was the only flying machine that even met the entry requirement of a round-trip mile-long flight. Although the aerial meet was clearly a failure, the world's fair as a whole was a success. Nineteen million people attended, and many of them enjoyed the electric wheelchairs that had been invented not to assist the disabled but to transport healthy individuals around the extensive grounds. The displays of new technologies were impressive, but what visitors remembered most was the introduction of the ice-cream cone. Santos-Dumont hesitantly tried one. It was not as light as a crepe, but he had to admit it was satisfying nonetheless.

[CHAPTER 13]

"AEROPLANE RAISED BY SMALL MOTOR, M. SANTOS-DUMONT PERFORMS A FEAT NEVER BEFORE ATTAINED IN EUROPE"

PARIS, 1906

•

SANTOS-DUMONT RETURNED to Paris with a fresh fondness for the city. However badly the Paris Aéro Club may have treated him was now a distant memory and an insignificant slight compared with the knifings his balloon had sustained in London and St. Louis. "As a youth," he recalled,

> I made my own first balloon ascension from Paris. In Paris I have found balloon-constructors, motor-makers, and machinists possessed not only of skill but of patience. In Paris I made all my first experiments. In Paris I won the Deutsch Prize in the first dirigible to do a task against a time limit. And now I have not only what I call my racing air-ship, but a little "run-about," in which to take my pleasure over the trees of the Bois, it is in Paris that I am enjoying my reward in it as—what I was once called reproachfully—an "aërostatic sportsman"!

The city was also the home of his closest friends. Of all the people he knew, Sem was the only one whom he allowed to enter his apartment unannounced. Santos-Dumont's servants and workmen treated Sem as if he were a member of the family. Once Sem came upon his friend huddled at his desk over plans for an eleventh flying machine. (There was a *No. 10*, an airship designed to carry ten people, but it never left the ground with more than one.) No gasbag was evident in *No. 11*; it looked like a monoplane. But Santos-Dumont turned the plans over before Sem could get a good look. On another occasion, Sem found him in his Neuilly workshop firing arrows from a cross-bow. He had replaced the arrow's feathers with cardboard wings of various shapes and sizes. He refused to discuss what he was doing but it was clear even to his nonscientific friend that he was studying the aerodynamics of wings in anticipation of building a plane. Sem fetched a bottle of champagne and the two of them celebrated without exchanging a word.

Santos-Dumont did not actually do much research in aer-odynamics. For all we know, his afternoon with the crossbow was largely the extent of it, unless one counts a few days of kite flying after he read an article about the Australian inventor Lawrence Hargrave's work on the lifting power of box kites. "I have never sat down to work seriously on abstract data," Santos-Dumont said. "I have performed my inventions through a series of tests, fortified by common sense and experience." The ultimate cowboy test pilot, he preferred going up in a novel contraption to doing laboratory research. This approach was the opposite of Langley's and the Wrights'. After days of twirling dead birds and prospective wing shapes on the arm of his turntable, Langley had built a series of progressively larger model planes before constructing the full-scale version and then sending someone else up in it. Santos-Dumont never thought of using a surrogate, although he had many volunteers. His personal ethics prevented him from putting someone else's life at risk. Moreover, he wanted to be the one to experience the thrill of flying an untested machine. The Wright brothers, for their part, conducted extensive experiments. When the wings of their gliders did not provide the lift that published tables suggested, they overturned established theory by constructing a wind tunnel and determining for themselves what the right wing shape was. Then they made close to a thousand glider trials before attempting powered flight.

Santos-Dumont skipped gliders altogether; he went directly from gasbags to planes. But he was not reckless. Before going up in his airplane, he put it through its paces in a way that neither Langley nor the Wrights could: He tested its stability by suspending it from one of his soaring airships. This eye-catching test was the first time most Parisians learned of his interest in heavier-than-air flight. The clandestine nature of this work was in sharp contrast to the openness of his experiments with airships, which were conducted to receive maximum attention. He never said why he had become so secretive, but it is fair to surmise. It was not because he planned to patent his new machine or otherwise profit from it. It was simply because he was a newcomer to heavier-than-air flight, and the only way he might overtake those who had worked on the problem for years was to catch them by surprise. When he had conducted his pioneering work with powered balloons, he was so far ahead of other aeronauts that there was little chance of someone circumnavigating the Eiffel Tower before he did; he could afford to reveal everything. But now, with the increased competition, he knew it was unlikely that he would be the first to fly in a heavier-than-air machine, and if he did not succeed, he did not want people to know that he had tried and failed.

At first he found the work more difficult than anything he had ever done with balloons. *No. 11,* whose plans Sem stumbled on, began life as an unmanned monoplane glider of questionable stability. Equipped with pontoons, it barely left the water when it was towed behind a speedboat. His sketches showed that he planned to add two engines to it to turn it into a twin-propeller aircraft, but he never did this because of the stability problem. He also did not succeed with *No. 12,* a helicopter with two propellers. The task of designing a suitably light and powerful engine for a helicopter was well beyond engineering know-how in 1905. Unaccustomed to failure, he returned briefly to his first love, the airship. He designed *No. 13,* a large "aerial yacht," and described his expectations for it in *Je sais tout:* "And if I were to tell you that I count on giving, this very summer, a new impulse to aerial navigation? That I even expect, before the termination of my experiments, to be able to cruise over Europe for a whole week in an aerial yacht which will have no need to come down at night because it will really be a flying house?" The secret to *No. 13*'s staying power was that it combined a hydrogen balloon with the heat source of a *montgolfière.* It was designed to stay aloft for days at a time because any hydrogen that seeped out could be replaced by hot air generated by the gas burner. That was the theory anyway. In practice *No. 13* might have gone up in a fireball. Santos-Dumont wisely canceled a planned test because even he had second thoughts about an open flame so close to hydrogen. His friends were relieved that he backed down, but his competitors mocked him for considering such a dangerous idea in the first place.

In August 1905, he completed another substantial airship, *No. 14,* and tested it away from Paris, on the Trouville beach near Deauville, a summer resort on the English Channel. Darting back and forth over the water, Santos-Dumont was pleased to find that he had as much control of *No. 14* as he had of his much smaller airships. When word got back to Paris about *No. 14,* his fellow aeronauts shook their heads. The world seemed to be passing Santos-Dumont by. He was stuck in the age of the plodding, pudgy, wind-fearing airship while they, impressed by recent successes with gliders, were sure the future lay in speedy, trim airplanes. But Gabriel Voisin, a talented twenty-five-year-old engineer whom Paris aeronauts were competing to employ, was pleased with the news from Deauville. He had made up his mind to join Santos-Dumont's team, and he knew that the Brazilian had a trump to play. *No. 14,* to be sure, was not a mundane airship but an aerial tug ready to be pressed into the service of heavier-than-air flight.

During the winter of 1905–1906, Voisin taught Santos-Dumont what he knew about airplanes. He shared stories of gliders that had flown and those that had crashed and killed their pilots. They talked about what it took to glide safely and how best to add a power source. Behind closed doors at Neuilly, they started building an ungainly looking airplane.

Santos-Dumont once again relied for power on an automobile engine and for lift on a long biplane wing made of box kites cobbled together with pine struts and piano wire. For an initial test of stability, he and Voisin hung the plane from a rope on a pulley and coaxed a reluctant donkey to pull the contraption. The designs of his rivals, while certainly more respectable-looking, did not seem to be performing any better. French engineer Louis Blériot's glider, pulled not by a donkey but a motor boat, plunged into the Seine as spectacularly as

Langley's machine had dived into the Potomac. And across the Channel, in London, Percy Sinclair Pilcher, a prominent meteorologist and amateur aeronaut, died when his test craft jackknifed. "All attempts at artificial aviation are not only dangerous to life," said the London *Times* in 1905, "but doomed to failure from an engineering standpoint."

On July 19, 1906, Santos-Dumont and Voisin retired the jackass and tested the prospective plane by dragging it from airship *No. 14,* the aerial tug. The strange spectacle satisfied Santos-Dumont that the box-kite design was stable. A month later, on August 23, he tried the plane *sans* balloon for the first time. He called it *14-bis* ("14-encore"). With a box kite positioned at the front of the plane and an engine in back, it looked as though it were flying backward. Because it had the shape of a bird with its box-kite "head"—what aeronauts called a canard configuration—the press dubbed it *Bird of Prey*. It was the first and perhaps only airplane in history in which the pilot had to stand the whole time.

The test was a success, and the *Herald*'s headline announced: "Aeroplane Raised by Small Motor, M. Santos-Dumont Performs a Feat Never Before Attained in Europe." At 5:00 A.M., Santos-Dumont set the fifty-horsepower motor in motion and the forty-foot-long machine with a thirty-three-foot wingspan darted forward along the ground at twelve miles per hour. It reached the end of the field, though, without taking off. He cut the motor and tinkered with it a bit, hoping to extract more power, before making a go of it back across the field. This time "every one present saw the wheels leave the ground and the machine travel forward, supported on air, six inches above the ground. This independent flight could not, however, be sustained long. The motor fluctuated, and thus what should have proved a steady upward movement in the air was changed into a series of gigantic hops." His fellow aeronauts were impressed by "the fact that a two-bladed propeller scarcely a métre in diameter can have a sufficient grip on the air to force the huge structure forward so rapidly along on uneven ground."

For the remainder of the day, Santos-Dumont tried unsuccessfully to improve the running of the motor. "I am more than satisfied," he said later. "I have accomplished more than I dared to hope. I am now certain that if the motor had been running its best I would have been able to sustain a prolonged flight along the surface of the ground. I am undecided how to act. I would like to mount a 50-horsepower motor in place of the one I have, but the makers say it will take eight days to arrange it and I am keen on continuing my experiments."

Once again he had his eye on a prize, actually two of them. Ernest Archdeacon, a lawyer and financier who was the new president of the Paris Aéro Club, had put up 3,500 francs for the world's first heavier-than-air flight of twenty-five meters, and the club itself was offering 1,500 francs for a trip of one hundred meters. Santos-Dumont summoned the judges on three different days. On the first occasion, September 13, 1906, *Bird of Prey* left the ground long enough for the judges to cheer before it stalled and headed straight toward them. They moved out of the way just as the plane made a heavy landing, smashing a propeller, the frame, and the wheels, but leaving Santos-Dumont free to hobble away. The eleven meters (thirty-seven feet) he traveled was too short for a prize. "But he flew," noted the *Herald*. "That fact was established beyond all possibility of contradiction. Although he had to return with a broken instrument, he had the satisfaction of knowing that he had accomplished before witnesses a feat never achieved in Europe by anyone but himself."

While his workmen spent the month repairing *Bird of Prey*, Santos-Dumont entered the Coupe Internationale des Aeronautes balloon race sponsored by the *Herald*. Given the progress in airships and heavier-than-air machines, the spherical balloon race was a throwback to an earlier era, but it captured the world's attention. A million people turned out at the Tuileries Gardens to watch sixteen balloons from seven nations ascend on the afternoon of September 30. There was something romantic about not knowing where they were going to land. Each contestant was given a card on which was printed in English, French, Russian, and Latin:

1—We have just come down in a balloon and are competing in a race for the Coupe Internationale des Aeronautes from Paris. Will you be so kind as to show us on this map where we are? 2—What country is this? 3—What is the name of the nearest important town and how far is it? 4—What is the name of the nearest railroad station and how far is it? 5—Can we obtain a cart to carry this balloon to the railway station? 6—Would you be so kind as to go and see if you can get us a cart? 7—Will you take me to the house of the Mayor or chief official of this place, as I wish to have this certificate of descent verified and signed by him in accordance with the rules of the race? 8—What is the name of this village or town, so that we shall not be taken by the Russian pollee or Cossacks for aerial anarchists?

As for provisions, the London *Tribune* reported that the three balloons from Great Britain carried "bottles of mulligatawny and mock turtle soup to be heated in caloret tins; essence of beef, pressed beef, pressed tongue and cheese, rolls, biscuits, sterilized milk, essence of coffee, bottles of mineral water, champagne, which is useful as a pick-me-up, and brandy, which is taken for medical action, should there be any faintness at a high altitude. For the same reason each balloon will be provided with cylinders of oxygen, so as to overcome the effects of rarification, should any inconvenience be experienced on that score." The Brits also took up several gallons of water "for washing purposes, and it has the advantage also of forming ballast." Another English paper worried that the ride would be choppy and predicted that "the great balloon race should be full of excitement for those whom fate chances—those who receive a bottle on the nape of the neck, an anchor in the small of the back or a leg of a chair in the pit of the stomach."

The competition was one of the rare times when Santos-Dumont did not win the prize. He had entered a spherical balloon that had two horizontal propellers extending from the basket. The idea was that they would control the vertical elevation, eliminating the need to carry ballast, freeing up valuable weight for additional provisions. But on the first day, the sleeve of his leather coat got caught in the mechanism controlling the propellers. He wrenched his arm and was forced to descend for medical treatment at Bernay, one hundred miles outside Paris. With his arm in a sling, he was back in his workshop three days later, helping his mechanics attend to the repair of *Bird of Prey*.

He tested the plane again on October 23, 1906. After a dozen failed starts and assorted snags, he signaled his workmen at 4:45 P.M. to release the machine. The plane charged across the field and gently climbed ten feet into the air. "The crowd was stirred to the utmost pitch of enthusiasm," the *Herald* reported. They watched *Bird of Prey* travel fifty meters or so, more than half the length of a football field, before it started to turn, its trajectory a "graceful curve." The spectators "expected to see a circle performed, but the motor had been stopped and a fall resulted." Still Santos-Dumont had flown twice the distance required for the Archdeacon prize, and he intended at first to fly even further. "I really do not know why I didn't go on," he said, as he extricated himself from the plane. "For an instant it seemed to me that the machine was making a sideward movement, and I foolishly cut off the gas. This all arises from inexperience. After a few trials I am certain I shall be able to travel many kilometers. The sensation was delightful. When the wheels left the ground I felt as if I were in a balloon propelled by some hidden force. I am absolutely certain that when I make a few alterations I shall be able to fly about with comparative ease."

On November 12, he was at it again, trying to double the length of his flight so that he could add the Aéro Club prize to the Archdeacon one. But this time he faced competition. Louis Blériot, who had also been helped by Voisin, was waiting in the Bois with his new biplane. The Aéro Club judges feared an altercation, with Santos-Dumont accusing Voisin of two-timing him, but in fact the Brazilian was so confident that he was solicitous of Blériot. After a few unsuccessful runs with a sputtering motor, Santos-Dumont encouraged the Frenchman to try for the prize. Blériot accepted the generous offer, but his plane collapsed without leaving the ground. As darkness set in, Santos-Dumont climbed into *Bird of Prey* for the fourth time and took off into the wind. The onlookers were so excited that they ran in front of the approaching plane. Santos-Dumont panicked. "He raised the head of his machine," the *Herald* reported, and soared upward until he was above the people, still traveling and rising all the time. Then the women beneath took fright and commenced to scatter and rush hither and thither. One or two women fell and the confusion was general. All this tended to unnerve M. Santos-Dumont, who hardly knew where to take his route. He endeavored to turn sharply to the right in order to win some clear space, but the movement was too brisk, and, fearing a heavy fall or a complete turnover, he cut the gas and descended. In touching the ground one of the wings was slightly damaged and a wheel bent.

According to his Cartier wristwatch, he had flown the 722 feet in 21.2 seconds. Despite the *Herald*'s enthusiastic description of his "rising all the time," he was never more than fifteen feet off the ground.

"I am very pleased," he told a bystander,

but am also very disappointed that I was prevented from completing a much greater distance by reason of the stupidity of the crowd, which in its eagerness to see everything came beneath my machine. When I saw the mass of people beneath me I confess I lost my nerve. I hardly knew what to do to avoid a serious disaster. I could hardly judge what clearance I had above the people's heads, and I decided hesitatingly to turn to the right. But for the moment my nerve was gone, and I felt the only thing was to descend as best I could.

The crowd stormed the plane, lifted him into the air, and for half a day carried him up and down the boulevards of Paris. He was quickly lionized around the world. "Santos-

Dumont Is Conqueror of the Air!" the newspapers exclaimed. Within a year, seven other aeronauts in Europe, inspired by his example, would follow him into the air in planes of their own design.

He flew *14-bis* only one more time. He knew the plane's shortcomings and moved on to other designs, but they were not successful. On March 27, 1907, he abandoned another biplane, *No. 15,* before it was airborne. As it raced along the grass to take off, it lost its balance. One wingtip brushed the ground and the plane collapsed, leaving him bruised and bloodied but not seriously hurt. *No. 16* was a winged airship; in June 1907, it too fell apart on the ground. *No. 17,* a biplane, was never built, and *No. 18,* a hydroplane, floated on the Seine but never left the water. By fall, he had assessed these failures and decided that he would try to do with the airplane what he had done so well with the airship—build the smallest possible machine that could safely carry him. The result was *No. 19,* a light and simple bamboo monoplane small enough to be easily transported. It was twenty-six feet long, with an eighteen-foot wingspan and a wooden propeller a yard in diameter. He had managed to reduce the fuselage to a single bamboo pole. At the back of the pole was the rudder, in the middle were the wings, and adjacent to them were the propeller and an eighteen-horsepower converted Dulthiel-Chalmers motorcycle engine.

If the seating arrangement in his airship was precarious, the one in *No. 19* was reckless. Three tricycle wheels extended down from the pole, and he sat on the axle between the front two wheels. Inches above him was the forty-nine-pound engine. If it spit fire or exploded, his head would suffer. There was also the danger that the engine would fall off the single pole and smash into him. Even if it stayed put, its position on the pole above him made the plane top-heavy and unstable. Santos-Dumont hoped that *No. 19* would earn him a new prize, the *Grand Prix d'Aviation*, which Henry Deutsch and Ernest Archdeacon had offered to the first plane to fly a round-trip course of one kilometer (six-tenths of a mile). But it was not to be. Although *No. 19* did get off the ground, its instability brought it down. Santos-Dumont retired the plane on November 21, 1907, after it was severely damaged on a four-hundred-foot hop in the Bois. (On January 13, 1908, Henri Farman won the *Grand Prix d'Aviation*.)

During the months that Santos-Dumont flew *14-bis*, the Wright brothers had been dismissive. After Santos-Dumont's 722-foot flight won him the Archdeacon prize, the Wrights told the Dayton press that his effort "does not appeal to us with the same degree of importance that it does to the people on the other side of the water, where the aeroplane is comparatively new in the problem of aerial navigation." But the brothers were ignorant of the details of the flight. When Chanute pressed them for further reaction, Wilbur said that he doubted the Brazilian could have flown more than a tenth of a kilometer. "If he has gone more than 300 feet, he has really done something; less than this is nothing." In fact, he had gone almost two and a half times that distance, and newspapers were calling on the Wrights to come forward and fly in public. "M. Santos-Dumont in a few months appears to have achieved more . . . than any other inventor," the *Herald* editorialized, "unless it be the Wright brothers, of Dayton, Ohio, who have surrounded their trials with secrecy and mystery."

In the definitive biography of the Wrights, Tom Crouch analyzed their peculiar indifference toward their rivals:

Between the time of Santos's short hop in the fall of 1906 and the first public flights of a Wright airplane in the high summer of 1908, a handful of European and American pioneers struggled into the air. Their aircraft were far more primitive than the Wright machine and the distances much shorter than those the Wrights could fly. Their activity was inspired by stories of the Wright success, and their machines were based on a sketchy understanding of Wright technology. None of that mattered. They had flown, and the whole world knew it.

The brothers took a strangely detached view—the European machines were much inferior to their aircraft; few of them incorporated any means of lateral control. None, by their definition, was a practical flying machine.

They were correct. Yet they lost something intangible by not making the first public flights. However superior their machine, Europeans saw their own colleagues fly at a time when the Wrights were still regarded as *bluffeurs*.

The Wrights had not envisaged that someone sufficiently daring might fly a considerable distance in a machine that could scarcely be controlled. That, in fact, was what occurred.

The control of a plane is achieved in three dimensions, corresponding to three axes of rotation: yaw, pitch, and roll. Pitch, or nose-up, nose-down control, is governed by the elevator, or horizontal rudder, which the Wrights put at the front of the plane, but which other aeronauts eventually placed behind the wings. The front elevator was the Wrights' answer to the problem that a decade earlier had killed Otto Lilienthal, the renowned German hang-glider pilot and aerodynamics theoretician. Lilienthal had made nearly two thousand successful glides, but on August 9, 1896, a sudden gust of wind stalled his monoplane glider and it fell fifty feet to the ground. It was the news of Lilienthal's death that drew the Wrights to aeronautics in the first place. By angling the plane's nose upward, the elevator could avert a fatal plunge in the event that the machine suddenly lost flying speed.

Yaw, or right-left control, is generally governed by the vertical rudder positioned at the rear of the machine (with the notable exception of the *14-bis,* in which the rudder was combined with the elevator in the front). All of the Wrights' rivals used rudders and elevators, but unlike the Wrights, they neither sufficiently appreciated nor solved the issue of controlling "roll," rotation about the plane's longitudinal axis, where one wingtip rises and the other falls. Langley had noticed that the wings of turkey vultures were angled slightly upward, and so he positioned the wings of the Aerodrome similarly. Setting the wings at a dihedral angle provided a measure of lateral stability in a straight flight but offered little protection against roll when the machine turned right or left. The Wrights, who also studied soaring birds, downplayed the importance of the dihedral configuration because not all birds employed it. But they noticed that soaring birds subtly flexed their wings rather then holding them rigid. Believing this to be the key to lateral stability, the Wrights invented the idea of "wing warping," twisting the wings so that the right one faced the wind at a different angle than the left one, so that the differential lift on the wings counteracts any tendency to roll. In the 1903 Flyer, in which the pilot lay prone between the biplane wings, he warped them by moving his hips in a cradle from which cords extended to the wingtips. At Kitty Hawk, the prone position of the pilot meant that on a hard landing he got a face full of sand, and so in subsequent Flyers the controls were reconfigured so that the pilot sat upright, but the crucial wing warping was maintained. (Nowadays wings are not twisted, but the same effect is achieved by the ailerons, or small flaps, on the rear edge of the wings.)

On *Bird of Prey*'s trial on October 23, 1906, what observers charitably characterized as the graceful curve of its flight path was in reality Santos-Dumont losing control of the plane's lateral stability. As the machine started to rock uncontrollably, he cut the engine and came down hard, the wheels collapsing on impact. By the next flight, on November 12, 1906, he and Voisin had added two octagonal surfaces—rudimentary ailerons—between the outer wing struts. Cords extending from these struts were sewn into the back of Santos-Dumont's jacket, or tied around his arms, so that the position of the surfaces could be adjusted by his swiveling his body, a method as novel as the Wrights' hip cradle but more difficult to operate. To adjust the ailerons, Santos-Dumont did what his contemporaries described as a kind of rumba dance. But it was not clear what the octagons actually accomplished, and on the November 12 flight, when Santos-Dumont turned to avoid the crowd, he again found the plane slipping sideways. He forced himself to descend sooner than he had intended not just because he was unnerved by the crowd but because he was again losing control of the plane. True, he had made the world's first public flights, but they certainly were not controlled ones.

Since the Wrights' secret test flights in 1903, they had perfected the Flyer in the course of more than a hundred flights at Huffman Prairie, a cow pasture eight miles east of Dayton, but they had had little success in selling their plane to a government. They expected the buyer to sign a purchase contract before viewing the plane and watching them fly. After Washington and London turned them down in 1905, they offered their plane to France for one million francs ($250,000). The

French, who thought the plane might be useful against their perennial enemy, Germany, were enthusiastic enough to put five thousand dollars in escrow toward the purchase, but in the end nothing came of it because the ministry of war was impressed by the flights of France's own aeronauts. The Wrights then approached Germany, which did not have a homegrown heavier-than-air program, but negotiations stalled because the kaiser's lieutenants wanted to see the plane before signing a contract. As biographer Crouch put it, "the Wrights proved to be almost as bad at business as they were good at invention."

Octave Chanute, one of the few people in whom the Wrights confided, urged them to wow potential buyers by flying triumphantly in public. They balked, and Chanute chastised them for letting their judgment be "warped by their desire for great wealth." The American press too had lost patience. In January 1906, *Scientific American,* which could hardly be accused of a French bias, questioned whether the experiments near Dayton had really taken place. The magazine addressed the rumor that in October 1905, Wilbur Wright had flown an astounding 24.2 miles in thirty-nine minutes. "Is it possible to believe," *Scientific American* wondered, "that the enterprising American reporter, who, it is well known, comes down the chimney when the door is locked in his face—even if he has to scale a fifteen-story skyscraper to do so—would not have ascertained all about them and published them long ago?" The *Herald* was even harsher, asking in an editorial on February 10, 1906, whether the Wrights were "fliers or liars."

The Wrights spent several months in Europe in 1906 trying to sell their plane but still they did not fly there. The few reports on the brothers that made their way into the European press were generally dismissive because of the lack of corroboration. And the reports themselves were riddled with errors, confusing the Wrights' Flyer with Langley's Aerodrome. The stories described the Flyer as being thrown into the air by a catapult rather than as taking off under its own power. This erroneous detail led French aeronauts, and later the Brazilian government, to maintain that even if the Wrights had gotten off the ground at Kitty Hawk in 1903, it was Santos-Dumont who deserved the real credit for having been the first to demonstrate *unassisted* powered flight in 1906. But the fact was that the Wrights had not used a catapult at Kitty Hawk; the Flyer took off on a sixty-foot rail so that it would not sink into the beach, but the rail did not propel it. On their first trial, a sixty-foot hop on December 14, the rail was inclined down a sand dune, so that gravity would help the Flyer pick up speed. But the Wrights deliberately laid the rail flat on the four trials on December 17, so that they could claim their place in history as the first to fly unassisted.

By September 1904, the Wrights, who felt they no longer had anything to prove, supplemented the rail with a derrick system. The rail itself was still necessary because the Flyer did not have wheels; it rode on two bicycle-wheel hubs, whose flanges hugged the rail. At Huffman Prairie, where the winds that helped the plane get airborne were not as strong as at Kitty Hawk, the Wrights found that they needed 240 feet of rail to build up sufficient speed. This often posed a problem: By the time they could cobble together a dozen twenty-foot lengths of rail and align them exactly, the wind would have shifted and they would need to take the rail apart and reassemble it in a different direction. By then, the wind might have changed again. They wanted to ascend from the original rail length of sixty feet, and their answer, in September 1904, was a twenty-foot derrick. A sixteen-hundred-pound weight, which was attached to the plane by a system of ropes and pulleys, would be hoisted to the top of the derrick. When the weight was dropped, it would yank the plane down the track. The system worked well, but fueled European skepticism about their claims to have flown unassisted.

By the end of 1907, though, their fellow countrymen at least were convinced of the validity of their claims; even *Scientific American* became a believer, having interviewed seventeen witnesses to their flights in North Carolina and Ohio. In 1908, the press-shy aeronauts finally decided that they had to win over their European doubters. Wilbur decided to stage a flight in Santos-Dumont's backyard. He arrived in Paris in late May and set up operations near Le Mans, a hundred miles southeast of Paris. Nothing at first was easy. The plane was badly damaged by heavy-handed French customs officials, who ripped the wing fabric and deformed the ribs, radiator, and seats. It took Wilbur ten weeks to repair the Flyer, and he had to recover from severe burns when an ill-fitting radiator hose sprayed him with boiling water. The French press, still portraying the Wright brothers as *bluffeurs,* caricatured Wilbur as an uncouth country bumpkin. He hated hotels, they said, preferring to sleep in a blanket beneath the wing of his plane. He bathed with a nearby hose. He ate from tin cans. His clothes had grease stains. He belched in public. He spurned red wine. He had none of *"l'elegance et l'espirit"* of Santos-Dumont and was a man of even fewer words. When asked why he was so reticent, Wilbur replied, "The only birds that talk much are parrots and parrots don't fly very high." Nancy Winters, the author of a picture book on Santos-Dumont, quipped that the Brazilian could have corrected Wilbur, pointing out that in his native land parrots flew extraordinarily high. But in fact Santos-Dumont never met either of the Wrights.

During the second week in August, Wilbur finally flew for the first time in Europe. On eight different trials he made tight circles over the Hunaudières racetrack, demonstrating to the world his supreme control of the plane. Back in the States, Orville was also flying in public, a prerequisite for winning an army contract. On September 11, 1908, he set an endurance record of seventy minutes and twenty-four seconds aloft at Fort Myer, Virginia. But things did not go well six days later, when Orville took up a passenger, Lieutenant Thomas Selfridge. On a flight over Arlington Cemetery, the right propeller broke, sending a blade slicing through the wires to the rudder. The plane barreled out of control toward the ground at fifty miles an hour. Orville was knocked unconscious, broke a leg and several ribs, and threw out his back. Selfridge fractured his skull and died on the operating table, the first fatality of an airplane.

The pressure was now on Wilbur to prove that flying was safe. Four days after Orville's accident, Wilbur broke his brother's record for time aloft, and by the end of 1908, he topped it with an astonishingly long flight of two hours and eighteen minutes. Even the most nationalistic Frenchmen now conceded that the Wright brothers were the true masters of the air, Santos-Dumont having stayed up for a mere third of a minute. The Wrights were the new favorites of Paris. The rich men who had surrounded the Brazilian now flocked to Wilbur's side. "Princes and millionaires," Wilbur wrote home to Orville, were as "thick as thieves."

French newspapers now praised aspects of Wilbur's personality that they had mocked only a short time before. "The smallest details of their lives seemed endlessly fascinating," Crouch wrote. "It was reported that the frying pan on which Will had done his cooking in the hangar at Camp d'Auvours would be displayed at the Louvre. Wilbur did most of his flying in a soft cloth cap that Orv had bought in France the year before. Now 'Veelbur Reet' caps appeared on heads all over France."

Santos-Dumont did not take this well. In public, he was gracious, declaring that the skies were big enough for everyone. In private, though, he was despondent. "It was, I may say now," he later wrote, "rather a painful experience for me to see—after all my work in dirigibles and heavier-than-air machines—the ingratitude of those who only a short time ago covered me in praise."

It was bad enough that the Wrights had usurped his position as first flier; what made it worse was his own recent failures to pilot five of his creations, *No. 15, No. 16, No. 17, No. 18,* and *No. 19*. He still believed in the idea of *No. 19*—a plane of minimal weight and elegant simplicity—and for a few weeks he skipped supper at Maxim's to work late perfecting an improved version, *No. 20*. He switched the position of the engine from above his head to under the seat. That cured *No. 20* of the top-heaviness that had plagued *No. 19*, but the engine's new position was no safer. It was "virtually in his lap, his legs straddling its hot manifolds and his toes but inches from the thrashing propeller belt." In March 1909, he unveiled *No. 20* in a meadow in St. Cyr, halfway between Paris and Versailles. The plane was as diminutive as *No. 19*, but more graceful. Its fetching silk-covered wings gave it the translucent elegance of a dragonfly, which inspired the nickname *Demoiselle* ("dragonfly" or "young lady"). *Demoiselle* was the world's first sports plane.

During the summer of 1909, he flew *Demoiselle* daily. Although it was too large to land in front of his apartment on the Champs-Elysées, it was the closest thing he had to an aerial car since *Baladeuse*. In *Demoiselle* he dropped in on friends at their country estates on the outskirts of Paris. One morning in September, he set out in *Demoiselle* to visit a fellow aeronaut five miles away in Buc. He took off from St. Cyr at 5:00 A.M. and reached Buc in five and a half minutes. His speed of 55.8 miles an hour was apparently faster than anyone had ever flown before.

He "became so casual about his aerial outings that he caused what must have been the first 'missing aircraft' alert," John Underwood wrote in his detailed account of *Demoiselle*, "The Gift of Alberto Santos-Dumont":

One rainy day in the latter part of September, he took off from St. Cyr and vanished among the dark clouds. After a couple of hours his apprehensive mechanic notified the authorities. The newspaper *Le Matin* responded by sending their ace reporter to the scene. Almost certainly Le Petit Santos had taken a fall, perhaps a fatal one. Then, at half-past one in the morning word was received from the Chateau d'Aion. The *Demoiselle* had alighted intact on the chateau lawn after completing an 18-kilometer (over 11 miles) flight in 16 minutes. Santos was, the caller continued, sleeping soundly, an overnight guest of the Comte de Gaillard,

whom he had not met before he landed in front of the count's castle.

In ballooning Santos-Dumont had recognized that it was safer to fly low, where he could take advantage of the guide rope. In airplanes it was the opposite. He had flown to Chateau d'Aion at an elevation of 650 feet. As he explained to the press,

This altitude was, as a matter of fact, necessary in the event of a crash-landing so that I might have enough time to find a place where my *Demoiselle* could land without the risk of being destroyed or getting damaged in a quick and mis-calculated return to the ground. . . . I state that the higher we fly, the farther we are from danger. Take poor Lefebvre for instance: when he died he was just 15 feet away from the ground, look at Bréguet, in Reims, who had just taken off when his craft was wrecked. The higher you rise, the more time you have to recognize danger. It's when you are nearer the ground that you should be more suspicious be-cause then, between the moment when you detect a stalling of a motor and the moment of landing, there is no time to take command of the situation.

I was right in taking precautions by having risen to such a height. My engine started misfiring; I deemed it advisable to take advantage of the odds which, suddenly, seemed to be in my favor: there was a castle lawn. Having no other choice I descended and gently landed, above all, so as not to cause damage to the lawn . . . I had entered as a tres-passer, not having come through the gate, without having my name announced. . . .

The Countess of Gaillard and her son . . . invited me to dine and when we were in the dining room, her other son stepped in and found me sitting where he thought I had died. He was coming from St. Cyr where he as well as other sportsmen had attended my departure and waited for my return in vain. I'm incapable of describing his bewil-derment.

"What about your airplane?" he asked.

It was in a stable like an ordinary horse.

The success of *Demoiselle* did little to elevate Santos-Dumont's spirits. His friends found that they could not console him. He accused them in turn of abandoning him, and whether or not that was true, Sem, Princess Isabel, and Aimé were no longer speaking to him. He whined incessantly about his diminutive size—an odd complaint given how well his lightness had worked to his advantage in aeronautics. He also told everyone that he was running out of money. Nobody believed him, but to humor him they urged him to patent *Demoiselle*. He refused. It was his gift to humanity, he said, and he would rather end up in the poorhouse than charge others for the privilege of copying his invention and taking to the skies.

With his blessing, the Paris automaker Clement-Bayard manufactured three hundred *Demoiselle*s, equipping the plane with one of their thirty-horsepower automobile engines and selling it in Europe for $1,250. They also opened a *Demoiselle* flight school, and Santos-Dumont was an occasional guest instructor. In the United States, Hamilton Aero Manufacturing, run by teenage inventor Tom Hamilton, sold the *Demoiselle* without an engine for $250, and a Chicago company offered a powered version for $1,000. *Popular Mechanics* magazine published Santos-Dumont's blueprints along with assembly instructions in its June and July issues in 1911, and "in a few months," Underwood wrote, "home-built *Demoiselles* were hopping criquet-like all across the country." Although *Demoiselle* was the first effort to provide the public with a personal flying machine, most people were too heavy to fly it. The pilot had to weigh less than 120 pounds, which accounted for the plane's popularity among high-school students:

Apart from bumps and bruises, no one was ever seriously injured in a *Demoiselle*, but catastrophes of the comic opera variety were common. Jean Roche, not yet 16, was appointed test pilot for a homemade specimen in 1910. It was assembled on the grounds of New York's Yankee Stadium. Roche recalls that the *Demoiselle* scooted over the ballpark at about 20 mph, bounded into the air with a jolt and landed in the outfield. The flight, if it can be termed as such, was precipitated more by contact with a ladder hidden in the high grass than by actual lift. Roche turned the craft around and taxied back to where the bystanders were gathered. Suddenly everyone was shouting and lunging forward with sand and buckets of water. Unbeknownst to Roche, the wheezing 14-hp Anzani had belched flame, igniting the wing. Half smothered, he tumbled out of the seat and fled. A moment later the *Demoiselle* went up in flames. The incident failed to dampen the young man's enthusiasm and fifteen years later he designed the first Aeronca [a lightweight monoplane, also known as the Flying Bathtub, that was affordable during the Depression].

Santos-Dumont himself had a serious accident in *Demoiselle* on January 4, 1910. The details are not clear because there were few witnesses and he did not speak or write about the mishap. According to one account, "a bracing wire snapped, collapsing a wing, and he plunged to the ground from 100 feet. Bruised and badly shaken, Santos declared that entanglement in the maze of wire had prevented him from being thrown out and killed. He had tumbled end over end three times coming down." That was the last time he ever piloted a plane. With no press and few adoring fans, his final flight was anticlimactic.

By the spring he was seriously ill. He suffered double vision and vertigo so severe that he dared not drive, let alone fly. When his symptoms did not recede, he consulted a doctor, who diagnosed him, at the age of thirty-six, with multiple sclerosis and told him that the disease would shorten his life. That night he boarded up his airship workshop and abruptly dismissed his loyal workmen. He retreated to his apartment for a week, refusing to see anybody. Word circulated that he had had a nervous breakdown. When he finally emerged, he claimed that he had been well all along. He told his friends that he was retiring early because he had done everything in life he had ever dreamed of doing. They played along but they could see that he was sick—although the diagnosis of multiple sclerosis was subsequently questioned by doctors who saw the problems as being chiefly in his mind.

Early in 1911, he slipped out of Paris and moved into a small house by the sea in Bénerville, near Deauville. Still too unsteady to pilot a plane, he kept in touch with the skies by taking up astronomy, mounting a fine Zeiss telescope on his roof. The house was relatively secluded and provided a welcome respite from the fervent nationalism in Paris, which escalated in the days before the Great War and viewed all foreigners with suspicion, even one as luminary as Santos-Dumont. On October 19, 1913, he emerged from his self-imposed exile to attend the unveiling of a monument to his aeronautical triumphs in Saint-Cloud. A huge statue of Icarus towered over a plaque commemorating both *14-bis* and *No. 6*, the Deutsch prize winner. Before a crowd of hundreds, rather than the thousands that had turned out in the past to see him fly, he spoke haltingly and awkwardly and thanked the mayor of Saint-Cloud for the honor. He handed the mayor a check for an undisclosed amount to be distributed to the poor in Saint-Cloud, and the mayor in turn congratulated him on his latest accolade, being made a commander of the French Legion of Honor. Santos-Dumont's fear of public speaking overcame him, and he left in the middle of the ceremony and returned to Bénerville.

The house in Bénerville was his principal residence for the next few years. He was there on August 3, 1914, enjoying his telescope, watching the stars at night and the sea during the day, when Germany declared war on France. He was horrified by the assault on his adopted country and intended to put himself at the service of the French military. But he did not get the chance—the French army reached him first. His neighbors had decided that the shy foreigner, who scanned the sea through a German-made telescope, was a spy for the kaiser. They imagined that he was signaling U-boats. When the military police raided his home, he was mortified. He had been the most famous and revered man in France, and now he was suspected of being a traitor. Although the police apologized after their search turned up nothing, the moment they left he set all of his aeronautical papers on fire. He burned every sketch, every blueprint, every congratulatory letter.

CHAPTER 14

" A W A R O F E N G I N E E R S

A N D C H E M I S T S "

•

THE SPECIFIC event that triggered the Great War—the assassination of the heir to the Austro-Hungarian throne by a
Serbian patriot—could not have been anticipated, although European leaders had expected hostilities to erupt. Military strat-
egists had long been planning for war. At the beginning of the
twentieth century, France still had not made its peace with
Germany's annexation of Alsace and Lorraine. Kaiser Wilhelm
II, jealous of the vast expanse of the British Empire, had
further territorial ambitions. And various regional conflicts
between other nations threatened to flare up. But military planners thought that war, when it came, would be of short
duration, anywhere from a week to a month, have limited
casualties, and involve only a limited number of countries fighting in a geographically restricted area. They expected an old-
fashioned conflict, a kind of group duel in which boys became
men by storybook acts of courage. No one envisaged a four-
year struggle that would engulf all of Europe and claim ten
million lives. There was nothing heroic about being sprayed
with mustard gas in a cramped, fetid trench and dying from a

burst of mechanized firepower unleashed by a distant anonymous foe.

At the end of the nineteenth century, the very forces of progress, which had done so much to enhance the quality of Western life, were fueling preparations for war. As Baron von der Goltz, the commander in chief of Turkish forces in Mesopotamia, observed as early as 1883, "all advances made by modern science and technical art are immediately applied to the abominable art of annihilating mankind." Many of the military applications were intentional: Breakthroughs in chemistry led to the creation of smokeless explosives, which increased visibility on the battlefield so that one side could see if it had really killed off the enemy. Other scientific developments had unforeseen consequences for the war effort: Advances in medicine, sanitation, refrigeration, and the purification of drinking water had decreased the infant mortality rate, but the burgeoning populations in Europe meant that there were now more young men who could be drafted into ever larger armies.

Technological utopians like Santos-Dumont lamented their failure to foresee a Europe-wide war and the omnibus military role that their beloved inventions would have. "I use a knife to slice gruyere," Santos-Dumont said in 1915. "But it can also be used to stab someone. I was a fool to be thinking only of the cheese." Until August 1914, many European intellectuals believed that no economically powerful nation would risk a major armed conflict because it would be too disruptive to international commerce. Inventions like the telephone and the railroad had required nations to cooperate in the setting of standards and protocols so that these technologies could operate across borders. The unprecedented level of cooperation, it was thought, would surely serve as a brake on military conflict. Up until the very start of hostilities, many of Santos-Dumont's fellow aeronauts were mastering Esperanto in the belief that this international language would soon displace French, German, English, Italian, and Russian.

Technology's greatest impact on the war, of course, was on the weapons themselves. Thanks to progress in metallurgy and machine-tooling, the decade following Von der Goltz's prophetic observation saw a five-fold increase in the range, accuracy, and firing rate of artillery and handheld guns. By 1893, according to one estimate, "infantry who had fired three rounds a minute at Waterloo could now fire sixteen rounds a minute." By 1914, the effective range of artillery had increased thirty-fold. "The famed Krupp works," wrote historian Michael Adas, "could produce massive 42-cm howitzers (the 'Big Berthas') capable of firing 1,800 pound shells as much as 10,000 yards in a trajectory that reached three miles at its apex." The United States' signature contribution to the Industrial Revolution had been the introduction of interchangeable parts and methods of mass production. All the belligerents in the war appropriated these factory methods—called the American system—so that they could supply millions of soldiers with the latest weapons and ammunition. As a result, in 1914, "a single regiment of field guns could deliver in one hour more firepower than had been unleashed by all the adversary powers in the Napoleonic wars."

An English physicist sadly observed in 1915 that the Great War was "a war of engineers and chemists quite as much as of soldiers"—a view developed in Adas's classic work *Machines as the Measure of Men:*

Railroads made it possible to move millions of soldiers into battle within days and—equally critical once the trench stalemate set in—reinforce points in the line where enemy forces threatened to break through. Wireless communications allowed staff generals and division commanders to coordinate the movements of tens or hundreds of thousands of troops over vast areas. . . . New techniques of food preservation and food canning made it possible to feed the huge national armies of recruits and conscripts over long periods of time, and mass production meant that they could be steadily supplied with helmets, uniforms, boots, and trenching shovels—which (ominously) the combatants of each of the great powers carried into battle from the opening days of the war. Once the war of maneuver sputtered to a halt after Schlieffen's grand design for the destruction of the French army had been frustrated at the Marne, barbed wire (which the Americans had invented to fence in cattle) and concrete and steel (which the Germans used the most ingeniously) were combined to build the massive field fortifications that would dominate the war in the west until the spring offensives of 1918.

In the decade and a half leading up to the war, there had been tepid efforts to reduce the effects of futile hostilities. In 1899, the Russian czar, Nicholas II, summoned twenty-six nations to an international peace conference at The Hague. Germany was a reluctant participant. Its leader, forty-year-old Kaiser Wilhelm II, was enthralled with all things military. His first public proclamation after he ascended to the throne in 1898 was not directed at the German people but at his soldiers. "We belong to each other," he said, "I and the army; we were born for each other." Like all commanders in chief, he expected the complete obedience of his troops, but he had a provocative way of asserting his authority. "If your Emperor commands you to do so," he told his soldiers, "you must fire on your father and mother."

Peace activists like Baroness von Suttner feared that Wilhelm II would set back their cause, and indeed a dispiriting tone was established on the first day of the Hague conference. The kaiser's representatives ruled out disarmament. There followed two weeks of fruitless discussion about limiting the growth of sea power: Britain, which had the largest naval force, supported any plan that locked in the status quo, while Germany, Japan, and the United States, which each had ambitions of controlling the oceans, spoke against a moratorium.

Surrounded by peace advocates from five continents, the delegates were embarrassed that they could not agree on the prohibition of a single weapon. Finally, by a vote of twenty-two countries to two, they outlawed dumdums, hollow bullets that expanded on impact. One of the two dissenting votes was cast by Britain, where the dumdums had been developed to stop the charge of "African savages" who refused to give up when they were shot with conventional bullets. The British delegate's impassioned dissent was supported only by the United States, which planned to test dumdums in the Philippines:

> The civilized soldier when shot recognizes that he is wounded and knows that the sooner he is attended to the sooner he will recover. He lies down on his stretcher and is taken off the field to his ambulance, where he is dressed or bandaged by his doctor or his Red Cross Society according to the prescribed rules of the game as laid down in the Geneva Convention.
>
> Your fanatical barbarian, similarly wounded, continues to rush on, spear or sword in hand; and before you have

had time to represent to him that his conduct is in flagrant violation of the understanding relative to the proper course for the wounded man to follow—he may have cut off your head.

By an even more lopsided vote, with the United States the lone dissenter, the conferees also banned the use of asphyxiating gas. The only unanimous vote came on a proposal to prohibit explosives and projectiles on balloons. The accord was easily reached because, in 1899, Santos-Dumont's efforts to turn unpowered balloons into flying machines were only beginning. Moreover, no country had yet successfully launched a bomb from a balloon, and none had plans to do so. Even Germany, where Count von Zeppelin had started working on mammoth rigid dirigibles, supported the proposal. Von Zeppelin was developing his gas giants with the idea of transporting civilians, although he was aware, from his days as a volunteer with the Union forces in the American Civil War, that balloons could be effectively employed for military reconnaissance. The ban on aerial bombardment was supposed to last indefinitely. Most of the countries were eager to confine hostilities to the ground and the sea. But two American colonels persuaded the other delegates to limit the ban to five years. They offered the same weapons-as-peacekeeper argument that Gatling, Nobel, and Langley had made, and they too would obviously be proven wrong by the course of future events. A time might come, they claimed, when explosives could be reliably launched from balloons. The mere threat of waging war from the air, they argued, would be so terrifying that countries would immediately lay down their arms and lives would be saved.

The five-year ban on air power expired in 1904, and the second peace conference at The Hague did not take place until 1907, but in the intervening three years no country had bombed another. At the second Hague conference, France opposed an extension of an outright ban. The French army had just purchased an airship from Paul and Pierre Lebaudy, two brothers who were inspired to take up aerostation after watching Santos-Dumont's flights but who had had none of the Brazilian's reservations about working closely with the military. What the French army intended to do with the airship *La Patrie* is not clear, but it wanted to leave its options open. The French representatives to The Hague ultimately agreed to a proposal that restricted aerial bombardment to military targets. The German army, concerned by the French stance and by news that *La Patrie* could carry enough fuel to penetrate eighty miles into Germany, pleaded with Von Zeppelin to build military airships.

At the time of the second Hague conference, Britain was indifferent to the potential military use of flying machines. Because the Royal Navy had dominated the seas for more than a century, the country thought of itself as immune to attack. That changed on July 25, 1909, when Louis Blériot flew the first heavier-than-air machine across the English Channel, claiming a five-thousand-dollar prize from the *Daily Mail* and redeeming himself for his earlier failure to beat Santos-Dumont into the air. The Brazilian sent his old rival a gracious note: "This transformation of geography is a victory for the air over the sea. One day, thanks to you, aviation will cross the Atlantic." Blériot responded: "I only followed and imitated you. For us aviators your name is a banner. You are our pathfinder." When the initial public excitement over the thirty-seven-minute flight abated, England was left feeling vulnerable. H. G. Wells, who the year before had written *The War in the Air,* in which a fleet of German airships attacks New York, was one of many who lamented: "England is no longer an island." The Royal Navy seemed powerless against airborne invaders. A cartoon in the press showed Napoleon's ghost asking Blériot, "Why not a hundred years earlier?"

In late August 1909, most of the world's aeronautical pioneers attended an aerial meet in Reims, France. Only Santos-Dumont, who was still distressed by the disaffection of his fans, and the Wright brothers, who indignantly insisted that they were not "circus performers," passed up the event. The Wrights' rival Glenn Curtiss, fellow bicycle mechanic, record-setting motorcycle racer (136.3 miles an hour), and the first American to fly after Kitty Hawk, was the top performer. A quarter of a million spectators watched him set a world speed record of 46.5 miles an hour on a twenty-kilometer course.

The rapid progress in flight was apparent even to skeptical military observers. Before 1909, "only ten men in the world had remained aloft for as long as one minute," Crouch wrote.

Eight months later, during the single week of flying at Reims, twenty-two airmen made one hundred twenty take-offs with twenty-three airplanes of ten different types. Eighty-seven of those flights were at least three miles in length; seven of them exceeded sixty miles. One pilot covered a distance of 111⅞ miles. The maximum altitude achieved was 508.5 feet; the top speed almost 48 miles per hour. All of the records set by the Wrights in the past year were shattered.

After Reims, airplane manufacturers sprang up across Europe and complacent militaries felt the need to purchase planes even if they had no idea how they were going to use them.

OF ALL technologically advanced nations, it was the United States, the home of the Wright brothers, in which aeronautical development seriously lagged in the prewar years. In 1909, the Wrights sold their first plane to the U.S. War Department for $25,000 plus a bonus of $5,000 if the speed topped forty miles an hour. The Wrights, financed by $1 million from old-money families like the Vanderbilts, formed a company to manufacture airplanes. Curtiss, capitalized at $360,000, started another firm. By 1910, ten other companies were selling planes and some fifty firms were providing parts and engines. But military and commercial markets were slow to materialize. Curtiss initially had no more success than the Wrights in interesting the military in a second plane, and in 1910, a forward-looking proposal to establish airmail died in Congress after newspapers like the *New York Telegraph* sniffed: "Love letters will be carried in a rose-pink aeroplane steered with Cupid's wings and operated by perfumed gasoline."

There were limited successes here and there. The first cargo flight took place in November 1910, when a department store in Columbus, Ohio, hired the Wrights' company to fetch a bolt of silk from Dayton. Although the Columbus *Journal* gushed over the flight ("The idea of a man flying up here from Dayton— where is your railroad train, your traction, your automobile now? Relegated to bygone days, along with the stage coach and the canal boat"), the cargo business did not take off. The current generation of planes lacked the space for large or heavy pay-loads, and it was prohibitively expensive to transport small loads. (The department store was able to recoup the cost by cut-

ting the silk into tiny pieces and selling them as aerial souvenirs.)
Similar experiments with airmail in 1911 demonstrated the
speed of delivery but not the cost-effectiveness.

In 1913, the first businessman commuted to work by plane.
Harold Foster McCormick built a hangar in front of his lake-
side house in Evanston, Illinois, and flew a seaplane twenty-
eight miles to the Chicago Yacht Club, near his office at
International Harvester, the family business. Alfred Lawson,
an aeronautical futurist, also commuted by seaplane, from his
home on Raritan Bay, in New Jersey, to his Manhattan office
near the East River. McCormick and Lawson did not start a
trend: At a time when a plane cost ten to fifteen times the
price of a car, which itself, at five hundred dollars, exceeded
the average worker's annual wages, most Americans could afford only a bicycle when it came to transportation.

The country's first regularly scheduled passenger flights,
from St. Petersburg to Tampa, began in 1914 on New Year's
Day, at the height of the Florida tourist season. A fleet of three
hydroplanes made four flights a day. A round-trip cost a steep
ten dollars, and the nineteen-mile trip took thirty minutes,
beating passenger boats by two and a half hours. That January,
184 people made the journey. The St. Petersburg–Tampa
route was the first passenger-plane service in the world (although zeppelins had been regularly ferrying far greater numbers of people in Germany).

By 1915, eight American airplane manufacturers had gone out
of business. The high cost of their planes and services had interfered with their success, but another obstacle was the fact that
most Americans did not yet trust the new form of transportation.
People refused to believe a plane could fly until they saw it for
themselves. As early as 1910, Curtiss and the Wrights had responded to the widespread disbelief by establishing exhibition

companies that toured the nation and put on air shows at carnivals and country fairs. That was also where the money was. The Wrights were charging as much as five thousand dollars a plane per show, and they paid their "birdmen" fifty dollars a day.

As Stanford University historian Joseph Corn documented in *The Winged Gospel,* for many people, watching a plane ascend was a quasi-religious experience, which could be described only with adjectives from the realm of the mystical: *miraculous, occult, inhuman.* At the first West Coast air show, in Los Angeles in January 1910, one observer wrote: "Thirty thousand eyes are on those rubber-tired wheels, waiting for the miraculous moment—historical for him who has not experienced it. Suddenly something happens to those whirling wheels—they slacken their speed, yet the vehicle advances more rapidly." Later that year the first plane flew over Chicago and a minister described the estimated million people watching: "Never have I seen such a look of wonder in the faces of a multitude. From the gray-haired man to the child, everyone seemed to feel that it was a new day in their lives." Some who watched the novel aircraft asked with all seriousness if they could hitch a ride to heaven.

Flying has always been associated with both theological extremes, the divine and the devil. Roman deities and Christian angels could fly, but so could witches, the devil's helpmates. Millions of Americans came to view planes as holy chariots, but their faith was severely tested by the number of flying machines that malfunctioned and sent their pilots spiraling to their deaths. The exhibition fliers were stunt pilots; they flew their planes in tight loops and upside down, often without the protection of seat belts. In 1912, Harriet Quimby, the first American female pilot and the first woman to fly across the English Channel, lost control of her plane at a Boston meet and died in a crash. Other female pilots achieved success in the early days of air shows, but Quimby's death—and the sheer expense of planes—dimmed egalitarian hopes that flying solo would empower American women, in the way that the bicycle had given an earlier generation of French women the freedom to travel beyond their villages.

Some pilots repeatedly pushed the safety envelope. Lincoln Beachey, a Curtiss pilot nicknamed the Flying Fool, commanded $1,000 to $1,500 a day for such perilous feats as flying into the misty gorge of Niagara Falls and under the suspension bridge spanning the river below. But he lived to tell about it, and survived numerous flights over farms where he tipped a wing close enough to the ground to stir up a dust cloud.

> Beachey challenged the concrete walls of Chicago's Michigan Boulevard, roaring down the street just above the heads of astonished pedestrians. At other times, Beachey felt that those watching were obligated to pay him for the privilege. At Ascot Park in Los Angeles, he noticed a group of citizens who had clustered in a tree to beat the admission charge. He banked around them with his usual alacrity, just having shaved the branches. A reproving press reported that the group of harried spectators sustained three broken arms and a skull fracture while making a precipitous escape. At Hammondsport, New York, in 1913, Beachey flew too close to a hangar roof and killed one spectator while injuring three.

And yet the press continued to portray him as a hero. In 1915, "he planned to fly a special plane he had built himself, which was intended to climb vertically. While he was testing it, the wings came off in a loop. A professional to the end, Beachey intuitively shut off the engine and closed the petcock to the fuel line before he fatally crashed."

In 1911, the Wright brothers had pulled out of the exhibition business because they had lost too many pilots: Four men had been signed to two-year contracts, and only one lived to fulfill it. Ill-conceived macho feats contributed to the fatalities, but the planes themselves were not the reliable machines that their manufacturers portrayed them to be. The planes could be controlled—until they fell apart. Whole wings broke off, guide wires snapped, wing-warping mechanisms jammed, engines stalled, wing fabric ripped, fuel tanks exploded, propeller blades fractured, and entire planes capsized in moderate gusts of wind.

In 1912, Santos-Dumont, sequestered in Bénerville, had a rare visitor who described the stunt flying in America, which was also being done in France but not as extensively. The world's first aerial showman was horrified. "I flew tight circles and steep turns because birds make them," Santos-Dumont told the visitor. "But show me a bird that makes loop-the-loops and flies upside-down. It's not natural."

The lurid spectacle of pilots plunging to their deaths may have slowed the development of commercial flight in the United States but it did not keep people away from air shows. In *Flight in America,* historian Roger Bilstein cited the words of a young army officer who attended the country's first international plane exhibition. "The crowd," he said of the tens of thousands of visitors to Belmont Park, New York, in October 1910, "gaped at the wonders, the exhibits of planes from home and abroad, secure in the knowledge that nowhere on earth, between now and suppertime, was there such a good chance of seeing somebody break his neck." The same ghoulish sentiment was expressed in a popular doggerel:

There was an old woman who lived in a hangar
She had many children who raised such a clangor
That some she gave poison, and some aeroplanes
And all of them died with terrible pains.

" C A V A L R Y   O F   T H E   C L O U D S "

•

WHEN THE Great War erupted in Europe in 1914, the military establishments of Germany, France, Great Britain, Italy, Russia, and Austria-Hungary possessed some seven hundred flying machines. Germany had the largest air fleet (264 planes and 7 zeppelins), followed by France (160 planes and 15 airships) and Britain (113 planes and 6 airships). All the countries had experimented with aerial bombardment before the war, but not one was satisfied with the results: The trial bombs usually missed their targets or failed to explode. And two actual cases of aerial warfare, in the Italo-Turkish War of 1911–12 and the Balkan War of 1912, proved the value of the plane as a scout but not as a bomber. Italy and Turkey had been fighting for control of Libya, and on October 23, 1911, Captain Carlo Piazza assessed Turkish troops in North Africa during an hour-long flight in the same kind of box-kite monoplane in which Blériot had crossed the Channel—it was the world's first reconnaissance flight over a war zone. Eight days later, a plane was used for the first time as an offensive weapon, when Lieutenant Giulio Gavotti flew one of the Italian army's six Blériot XI's over Libya, pulled out the pins of four softball-size grenades with his teeth, and lobbed one at the town of Ain Zara and three at the Taguira oasis, where Piazza had spotted Turkish strongholds. By the end of the Italo-Turkish War in October 1912, the Italians had flown 127 missions—some of them in airships—over Libya and released 330 bombs. In the Balkan War, the unfortunate Turks were again bombarded, this time by ten-kilogram explosives from Bulgarian planes. (Each bomb, which hung from a rope that extended from the plane's tail and looped around the pilot's foot, was released by kicking the rope free.) But in both wars, the planes and airships apparently caused no casualties and only minimal damage.

At the onset of the Great War, the combatant nations were enthusiastic about the use of flying machines for reconnaissance, and planes quickly established themselves as crucial elements in each country's arsenal. It was the French who first demonstrated the value of surveillance from the air: On September 3, 1914, Blériot planes detected a widening gap between the German First and Second Armies moving toward the Marne, and this knowledge enabled the Allies to halt the German advance. Aerial scouting made it difficult for each side to move infantry and artillery around without the enemy knowing it. Planes and airships also proved indispensable in directing firepower on the ground.

At first the airmen, in their one- and two-seater reconnaissance planes, were unarmed. In fact, when German and French fliers passed in the air, they saluted each other. Within the opening days of the war, however, one pilot—it was not clear who had this infamous distinction—brought along a pistol or a rifle and took a potshot at another plane. Soon every airman carried automatic weapons, and a technological race ensued to equip the planes with ever more lethal firepower. One problem was the propeller, which got in the way of a pilot firing straight ahead without destroying his own plane. Roland Garros, a French stunt pilot turned military airman, solved the problem by mounting steel plates on his wooden propeller blades. When he fired a machine gun in the direction of the propeller, the plates harmlessly deflected many of the bullets but let enough through to shoot down German planes. Garros's simple solution did not remain a secret for long. On April 18, 1915, engine failure forced him to land behind enemy lines. The Germans captured him, and while they gave him wine, beer, grilled meats, and pastries, their engineers dissected his invention. Within two days Anthony Fokker, who would manufacture a line of German combat planes, had improved on Garros. He was able to synchronize the propeller's rotation and the machine gun's automatic firing so that the entire spray of bullets passed between the whirling blades.

As the war progressed, tales of dogfights between intrepid fliers delighted a public numbed by the slaughter of millions of anonymous soldiers in the trenches. The horror of the mechanized warfare on the ground was captured by a chilling statistic, cited by Adas: In 1916, one thousand shells pummeled each square meter of the contested expanse near the French fortress of Verdun. And the shells that went off, he added,

sent shards of metal ripping through human flesh. As a nurse at the front, Vera Britain witnessed firsthand the effects of the handiwork of the engineer and the chemist. She treated "men without faces, without eyes, without limbs, men almost disembowelled, men with hideous truncated stumps of bodies." Shells that released chlorine, phosgene, or mustard gas left their victims "burnt and blistered all over . . . with blind eyes . . . all sticky and stuck together, and always fighting for breath."

By contrast, the clash of aerial gladiators, whose destinies depended on their own ingenuity, and whose names and life stories were recited in the newspapers, was a welcome throw back to the days when war itself seemed noble. "It is a youth-intoxicated profession, aviation," a pilot in the Royal Navy wrote in 1918. "It is a calling pertaining exclusively to the cool daring, the iron nerve, the reckless abandon of youth. Perhaps a little of the glamour and mock heroism of the early days has worn off. But the airmen remain the darlings of the gods."

The courtly treatment that the Germans afforded Garros was typical of the respect all the combatant nations showed toward "men from the sky." As the Royal Navy pilot explained,

The "bon camaraderie" that has developed between the enemy service and our own dates back to the days when a British airman was earthed in their country by shrapnel, and his squadron informed of the unfortunate one's plight by a note dropped from an enemy raiding machine. Since that time the custom has never lapsed. The appreciation was mutual, as evidenced by our pilots, and an incident that recently occurred in Berlin. There, at an exhibition of "Air War Booty," a department was devoted to [Oswald] Boelke, renowned Fokker pilot, and a wreath of violets was on show, from the pilots of the R.F.C. [Royal Flying Corps], "In memory of a gallant and noble enemy."

When Manfred von Richthofen, the infamous Red Baron who shot down a record eighty Allied planes, was finally killed, the British buried him with "military honours in unstinting recognition of his great courage, his sportsmanship, and his tireless, relentless spirit." An Englishman whose compatriots died at Richthofen's hands was nonetheless so full of praise for him that the words sound like they were written by the Red Baron's own family:

> He fought, not with hate, but with love for fighting. It was his joy, his sport, his passion. To him, to dare and to die was to live. He had the courage to kill and be killed. . . . He was courageous, and knew it, gloried in it, flaunted it with his challenge to the world of his enemies. He made them know him—he put his name on their lips—his name that was unknown, unheard of, when he started the war as a second "looie." Wounded and decorated, he became the guest of kings and queens. Boys and the youth of a nation made him their idol, cheered him, followed him on the street. He was young and blond, shy and handsome, proud and serious. Girls by the thousands worshipped his picture and filled his mail with letters by the sackful. One of them he loved. He wanted to make her his wife, but he did not want to make her his widow. He knew he was going to be killed. He won the admiration and respect of his enemy.

Those who glorified Richthofen and other airmen ignored the fact that their deaths were gruesome. "Some went down like flaming comets, burned beyond recognition before the charred remains struck the earth thousands of feet below," an observer noted. "Others plunged earthward through the blue in drunken staggers as their bullet-riddled bodies slumped forward lifelessly on the controls. Some fell free from shattered planes at fearsome heights, poured out like the contents of a burst paper bag, and some, hurtling down in formless wrecks, buried themselves in the ground." Being a member of an elite air corps, where one slept at night in a warm bed instead of in a trench alongside decaying corpses, was hardly a guarantee that one would make it through the war. Fifteen thousand British, German, and French airmen died by the armistice. Another seven thousand were missing or imprisoned, and seventeen thousand were wounded. These numbers were dwarfed by the carnage on the ground, but the airman's job was a dangerous one: The odds were better than fifty-fifty that he would end up a casualty.

Even when the airman's duties expanded to include bombing sorties, his reputation did not diminish. For the British prime minister, Lloyd George, the fliers remained the "cavalry of the clouds":

> They skim like armed swallows, hanging over trenches full of armed men, wrecking convoys, scattering infantry, attacking battalions on the march. . . . They are the knighthood of the war, without fear and without reproach. They recall the old legends of chivalry, not merely the daring of their exploits, but by the nobility of their spirit, and amongst the multitude of heroes, let us think of the chivalry of the air.

The bombing started almost as soon as the war began. When Germany declared war on France on August 3, 1914, one of the provocations it cited was the supposed French bombing of Nuremberg the day before. The event never took place—Nuremberg was too far from the border for French planes to have reached it—but Germany retaliated anyway. On August 6, zeppelins bombarded French forts at Liège that were impeding the advance of German troops. And by the end of the month, a Taube monoplane tossed five small bombs at a Paris railway station. They fell wide of the target but killed a woman at 39 rue des Vinaigriers. She was, according to military historian Lee Kennett, "the first of some five hundred Parisians to die as a result of German air and land bombardments. Yet there was something almost quixotic about the attack. Along with the bombs, the German pilot threw out a message attached to a streamer in the colors of the German Empire: 'The German Army is at the gates of Paris. You have no choice but to surrender. Lieutenant von Hiddessen.' " Because the Paris railroad carried French soldiers to the front, Von Hiddessen's bombing raid was arguably not in violation of the Hague ban on attacking civilian targets. Some fifty grenade-size bombs rained down on Paris before the end of 1914. They did little damage except to French pride when one of them landed on Notre Dame.

On January 7, 1915, British bombers targeted a railroad station in Freiburg, Germany, but, as was typical, the small incendiary devices fell far afield. A number of civilians were killed, and the outraged kaiser, who had so far spared England because it was the home of his relatives and friends, ordered zeppelins to attack the country. At this point in the war, the mammoth airships posed a greater danger than planes because they were large enough to transport hundreds of pounds of bombs. On January 19, 1915, two zeppelins attacked Yarmouth and King's Lynn on the coast of England. They left Germany at dusk and arrived at night, identifying the target cities by their streetlights. They were back home by dawn. In the spring, the kaiser ordered the zeppelins to attack London but to stay clear of its palaces. On May 31, 1915, the 536-foot-long LZ-38 dropped forty-four hundred pounds of explosives on the British capital, killing twenty-eight people

and wounding another sixty-one. The British eventually fig-
ured out how to shoot the zeppelins down. They were easy
targets because of the highly explosive nature of the hydrogen
gas. Still, the airships managed to kill more than five hundred
men, women, and children in London alone.

In May 1917, the Germans attacked London again, this
time in aircraft, two-engine Gothas and multiengine R-planes,
each capable of carrying one ton of bombs. The planes struck
the city twenty-seven times over the next year. "In grim statistical terms," Kennett concluded,

> the new bombing planes were more efficient than the zep-
> pelins had been. They took more lives with a smaller
> amount of explosives. But when the effects of all the attacks
> on England, by zeppelins and planes alike, are added up,
> the figures are surprisingly small. The Germans dropped
> on England less than 300 tons of bombs, killing fourteen
> hundred people and injuring forty-eight hundred more.
> These are the kinds of totals one would find reported for
> a single "quiet" day on the western front. Total property
> damage was put at slightly over £2 million—less than half
> what the Great War cost the British each day.

As a means of psychological warfare, the air raids were successful: They terrorized the civilian population. During zeppelin attacks, upper-crust Londoners who would not deign to
ride the subway found themselves scurrying in their night-
clothes to take shelter in underground train stations.

The destruction that Allied bombers inflicted on Germany
was comparable. In 1918, twelve hundred Germans were killed
or injured in 657 air raids. It was only shortly before the
armistice, on November 11, 1918, that the bombers on either

side were accurate enough to guide explosives into the narrow trenches that defined the Great War. All in all, fewer than 1 percent of the war's total casualties can be attributed to aerial attacks. The armistice came just in time, before the Allies had a chance to drop poison gas on German towns and before the Germans could lob their new 3,000-degree-Fahrenheit firebombs at Paris and burn, according to their calculations, a third of the city. After the armistice, Orville Wright was the latest in the long series of distinguished thinkers who convinced themselves that weapons would ensure peace: "The Aeroplane," he wrote, "has made war so terrible that I do not believe any country will again care to start a war."

[C H A P T E R   1 6]

D E P A R T U R E

G U A R U J Á ,   1 9 3 2

•

SANTOS-DUMONT SPENT most of the war years in Brazil. He attended aeronautical conferences in Washington and in Santiago, and he gave a few press interviews in which he tried to prove his case that he was the true inventor of the airplane. He reminded everyone that no official witnesses, no delegates from one of the world's prestigious aero clubs, had seen the Wright brothers fly before he left the ground in *14-bis*. In the midst of the war, nobody paid much attention to his argument. He also wrote a thin polemic, *O Que eu Vi, O Que nos Veremos* ("What I Saw, What We Shall See"), justifying his career. Both his writing and his spoken remarks were sullen and jumbled. He exhibited none of his old flair and joie de vivre, and he distorted history. For instance, he gave the wrong date for the Deutsch prize and claimed that his debut flight in his first airship occurred in a February snowstorm when it had in fact happened on a calm, late summer day.

On top of whatever physical problems may have exacerbated his mental illness, he was now terribly disheartened by the military use of flying machines. He was certainly not a pacifist. More than once he had offered his airships to the

French military. But he never imagined the carnage planes and zeppelins would cause. The bombing raids from airships—"my babies," he called them—particularly disturbed him, and he blamed himself for their invention. He held himself personally responsible for every fatality caused by a flying machine, and, to punish himself, he read as much as he could about the gory details of the deaths. "He now believes that he is more infamous than the devil," wrote an acquaintance, Martin du Gard. "A feeling of repentance invades him and leaves him in a flood of tears."

Santos-Dumont lived a decade and a half beyond the war's end but rarely had a moment of serenity. He traveled frequently between Europe and two homes in Brazil, his childhood house in Cabungu, which the Brazilian government gave him as a gift in 1918, and a small house called *La Encantada* ("the Enchanted Place"), which he built in the hills above Rio in Petrópolis, once the playground of the Portuguese royal family. Always the inventor, and still superstitious, he constructed an odd staircase in *La Encantada*. The left half of the first step was purposely cut away so that he and his visitors were forced to ascend with the right foot first. The top step was similarly constructed so that anyone going down the stairs also had to lead with the right foot. *La Encantada* was sparse—the adjoining maid's house was larger and posher—and he slept on a thin mattress that he positioned on top of a cabinet. He ordered all his meals from a nearby hotel and ate them in the house by himself. He occasionally ventured out to play pickup games of tennis at the hotel's court but he was such a bad loser—and he lost often because of his shaky motor control—that he would storm off without so much as saying good-bye to his opponent. He avoided visitors in Petrópolis. When the president of Brazil called on him, he refused to answer the door. At Cabungu he raised orchids and cattle imported from Holland, but he never settled down, there or in Petrópolis.

In the early 1920s, he lobbied most of the governments in Europe as well as those in North and South America to de-militarize flying machines. He appealed, too, to the League of Nations, saying, "Those, who like myself were the humble pioneers in the conquest of the air, had in mind more the creation of new means of peaceful expansion of the peoples of the earth than furnishing them with new methods of destruc-tion." He was politely received, but no country disarmed its aircraft.

One day he stopped eating. His relatives persuaded him to institutionalize himself but kept up the public ruse that he was doing just fine. Throughout the 1920s he checked himself into various sanatoriums in Switzerland and France, usually in re-mote towns where he was less likely to learn of disturbing developments in aviation. He spent the days binding books of poetry and for a brief time resumed his interest in aeronautics. He glued feathers to his arms and strapped on wings powered by a small motor in a backpack. A psychiatric nurse stopped him from jumping out the window to test his wings.

He tried to make something useful of the motor, however, by hooking it to his skis—there were no ski lifts at the time—to assist him in walking up the snow-covered slopes of St. Moritz. It is hard to believe that the motor was strong enough to be of much help, and although posed photographs exist of friends wearing it on their backs, there were no witnesses to his ac-tually using it to ascend a mountain. It is ironic that an aid to walking, the oldest form of conveyance known to man, was one of the last inventions of someone who devoted his life to leaving the ground.

He also invented a slingshot that could throw a life pre-server to a drowning victim, and he patented (apparently the only patent he ever took out) what might later be thought of as a Rube Goldberg–like contraption for greyhound racing that kept the dogs moving around the track by dragging a treat in front of them. Neither device was apparently ever used.

In December 1926, an acquaintance named A. Camillo de Oliveira visited him in the Clinique Valmont, in Glion-sur-Montreux, Switzerland. The aeronaut was in good spirits. He showed de Oliveira his hand-bound books, and the two planned to go skiing the next day. But when de Oliveira woke the next morning, he found a message from Santos-Dumont: "Dear friend, I couldn't sleep last night just thinking of skiing. I'll see you later. We'll arrange another one. I just don't have the strength to go skiing."

The year 1926 had been a disappointment for Santos-Dumont. While visiting France, he made an unannounced visit to the Issy-les-Moulineaux office of his *14-bis* collaborator Gabriel Voisin. Santos-Dumont was silent and fidgety. He did not respond when Voisin asked him how he was doing. Suddenly he pronounced his love for Voisin's seventeen-year-old daughter Janine and asked permission to marry her. Voisin was dumbstruck. Santos-Dumont had never shown any interest in women, and he barely knew Janine. But Voisin did not want to hurt his fragile friend's feelings. He explained that marriage was impossible given their thirty-six-year difference in age.

In August 1927, de Oliveira returned to Valmont, where the head of the clinic, a Dr. Wittmer, unexpectedly sought him out. "We know you have good relations with Mr. Santos," Wittmer said. "You're probably the only person in the position of giving him some advice without offending him. Even though

we enjoy his presence here, we believe it's our duty to ac-
knowledge that his long retreat here may cause him harm in
the long run. He may start fearing life in the 'outside' world.
After all, he's not disabled or incapacitated. Later on, if he
wants to come back, he'll always be welcome." Indeed, he was
already afraid of rejoining society. Three months earlier he had
passed up an invitation from Charles Lindbergh to join him
at a dinner in Paris to celebrate his historic flight across the
Atlantic. Santos-Dumont wept when he received the invitation
but politely declined, faking a prior commitment.

At Wittmer's behest, de Oliveira contacted one of Santos-
Dumont's relatives, his nephew Jorge Dumont Villares, who
came and retrieved his uncle from the clinic. By the end of
1928, Santos-Dumont felt well enough to return to Brazil. As
he sailed into the Bay of Rio on December 3, a dozen of the
country's top scientists and intellectuals boarded a hydroplane
christened *Santos-Dumont* and flew out to greet him. He stood
on the deck, smiling, delighted that his countrymen still remembered him. As the hydroplane dipped to drop balloons
and confetti, it exploded, killing everyone on board, setting
back not only Santos-Dumont's recovery but Brazilian science
itself. "I have always requested you not to fly upon my arrival,"
he told the organizers of the festivities. "The commotion
brings forth considerable carelessness. What a lot of lives sacrificed for my humble person!" He checked into a Copacabana
hotel and committed to memory all of the obituaries. Over the
next week, he attended all twelve funerals.

Eight years earlier, in 1920, he had helped workmen dig
his own grave at the Ceméterio de São João Batista in Rio.
He had insisted on removing the dirt himself. He commissioned a replica of the statue of Icarus at Saint-Cloud and

arranged to transfer his parents' remains to the tomb, leaving a place between them for his own body. In 1928, after the hydroplane tragedy, he visited the grave-in-waiting and ran his fingers through the soil. His thoughts kept returning to the crash, he told friends, and he decided to go back to Europe where there were fewer reminders of the accident. He sought refuge again in a sanatorium.

There was a brief period in the mid-1920s when it looked as if Santos-Dumont's dream of a universal personal flying machine—"an airplane in every garage"—might come true, thanks to the efforts of the man who was responsible for putting a car in virtually every American garage. In the first years of the twentieth century, the tremendous expense of a car had made the prospect of most families ever owning one seem remote. In fact in 1906, Woodrow Wilson, then the president of Princeton University, went so far as to suggest that a resentful carless underclass might foment a social revolution. Wilson's fear was alleviated in 1908, when Henry Ford introduced the Model T, nicknamed the flivver or tin lizzie, and through innovations in assembly-line production was able to drop the price so that ordinary people could afford them. By 1924 Ford had sold fifteen million flivvers. He was also in the airplane business, and in 1925 he started selling expensive eight- and twelve-seat commercial planes. The public clamored for him to produce an inexpensive "air lizzie." Aeronautical enthusiasts spoke of the salutary benefits of "driving" at high altitudes, where the air is pure, and social planners foresaw a mass migration to mountain and seaside towns from which the average worker would commute by air car. In 1926, Ford showed off a prototype of a one-person "Ford flying flivver," and the press, as historian Joseph Corn noted, was exuberant:

A New York *Evening Sun* columnist, imagining himself already aloft in the little machine, penned what he termed the "hallucination" of a new flivver owner:

> I dreamed I was an angel
> And with the angels soared
> But I was simply touring
> The heavens in a Ford.

Outside the big city, on the farms and in the workshops where Ford, himself a country boy, had his most devoted following, many took Ford's effort as prophetic. One country journalist, forgetting how skittishly farm animals had reacted to the automobile, claimed that "when Mr. Farmer lands in the farm yard" in his family plane, his chickens "will be there to welcome him."

Santos-Dumont was encouraged by news of aerial cars but his enthusiasm—and everyone else's—was short-lived. On February 25, 1928, test pilot Harry Brooks was killed in Miami when his flying flivver, only the third that Ford had built, crashed into the beach. A grieving Ford pulled out of the aircraft business altogether. The dream of the personal flying machine was subsequently kept alive only by fantasy illustrations—cheerful men pushing "helicopter coupes" into their suburban garages—on the covers of magazines like *Popular Science*. In reality, no technologist could figure out how to make a plane that was simple enough for the average person to operate. Efforts to simplify the plane's controls invariably made flying less safe—all the "complex" controls were there for a reason. And there was also the troubling issue, if the skies were ever populated with air cars, of midair collisions. Even today, with all the advances in aviation technology, Santos-

Dumont remains the only person in history to have fulfilled Jules Verne's vision of freedom in the air. No one else has enjoyed the convenience of a flying car that he had with *Baladeuse*, the little airship whose reins he would hand to the doorman at Maxim's and to the stable boy at the polo grounds in the Bois.

In 1929, Santos-Dumont wrote a short unpublished manuscript called *L'homme Mecanique* ("The Mechanical Man") and dedicated it to "posterity." It was a two-part document. The second part was a brief review of early aeronautics, a recapitulation of familiar arguments that he was the first to fly an airplane:

> The partisans of the Wright brothers assert that it is they who flew in North America from 1903 to 1908. These flights were supposed to have taken place near Dayton, in a field along a street-car line. I can't help being profoundly astonished by this ridiculous claim. It is inexplicable that the Wright brothers could have made innumerable flights for three and a half years without a single journalist from the perspicacious American press taking the trouble to see them and producing the best reporting of the time.

The beginning of the document was more revealing. It was a rambling, technical discussion of what he called the Martian Transformer, his invention for climbing mountains with skis. He was proud that he had found a way of transforming the rotary motion of an ultrapowerful backpack-size motor into the reciprocating motion of the skis. The name of the invention was a bow to H. G. Wells's *The War of the Worlds,* in which the Martians, ignorant of the wheel, "had automatic legs on all of their machines, including the immense war chariots that ravaged London." He believed that the invention was more important than *14-bis* or *Demoiselle*. In *L'homme Mecanique* he explained that he was going to use the same principle of converting rotary motion into reciprocating motion to build a set of feathered wings that a man could strap to his arms. Driven by a lightweight motor tied to his back, the wings would flap rapidly and carry him into the air. It would be the culmination of his own quest that began with *Baladeuse* and *Demoiselle:* the creation of the smallest possible, least cumbersome personal flying machine. In this case there was barely any machine, and no fuselage or frame to protect the man from the elements or the ground if he crashed. It was a thoroughly impractical idea but one that strongly appealed to his romantic temperament. It was a throwback to the Middle Ages, when winged men jumped to their deaths. If he succeeded in giving men wings, providing men with the ultimate freedom of movement, he felt that his aeronautical legacy would be secure. He never had the energy, though, to pursue the idea beyond his aborted effort in the sanatorium.

In 1930, he was disturbed by reports from his homeland that a revolution had broken out on October 3. His old acquaintance Antônio Prado was imprisoned, and Santos-Dumont wrote to Prado's wife: "I fell ill with the news from Brazil that I am afraid of becoming insane. I am in a private hospital."

In the days when Santos-Dumont was flying, he repeatedly evaded death. Now he confronted it wherever he turned. The day after his countrymen started killing each other, a British dirigible called *R101*, which had been bound for India in a highly publicized flight to demonstrate the safety of airship travel, crashed into a hillside in Beauvais, France, killing forty-eight passengers. Santos-Dumont needed to be restrained from hurting himself.

In 1931, his nephew Jorge retrieved him again, from a rest home in Biarritz, and the two of them returned to Brazil. The following year, a new war erupted, the so-called Constitutionalist Revolution, pitting pro-democracy insurgents from the state of São Paulo against the increasingly dictatorial president, Getúlio Vargas. At first Jorge and his uncle were living in the city of São Paulo, but Santos-Dumont's doctor encouraged him to move to a more tranquil place. Jorge arranged for his uncle to stay in a hotel in the beach resort of Guarujá. The two of them ate breakfast and lunch together in the hotel dining room. Other guests who recognized the aeronaut often came over to greet him, but Jorge intercepted them and explained that he was recovering from an illness and needed to eat in peace. Aside from his nephew, Santos-Dumont spoke at length only to the children who collected shells on the beach outside the hotel. He had given up on his appearance. No longer a fashion trendsetter, he was disheveled. He refused to wear a suit in the hotel. Formal dress was required at dinner, and so he and Jorge had room service every evening.

Jorge rose early each day and bowdlerized the morning paper to keep his uncle from learning that federal troops were bombing the Paulistas. But the news could not be kept from him for long. On July 23, 1932, when he was in the hotel lobby, he heard a plane fly past and bomb a nearby target. Feigning indifference, he sent his nephew on an errand and took an elevator back to his suite. Sixty-eight years later, the elevator operator, Olympio Peres Munhóz, still remembered Santos-Dumont's anguished words as he left the elevator: "I never thought that my invention would cause bloodshed between brothers. What have I done?"

He retired to his bedroom. He changed into his first suit

in months. He rummaged through the closet until he found two bright-red ties from his flying days in Paris. He knotted them around his neck, picked up a chair, and headed into the bathroom. His nephew, who had doubts about leaving him alone, returned too late. Santos-Dumont, age fifty-nine, was dangling at the end of the ties from a hook on the bathroom door—a method of suicide that could have succeeded only for someone so light.

The local police sealed off the hotel room, and, acting on orders from as high perhaps as Vargas, declared that he had died of cardiac arrest. The coroner faked the death certificate. When word of his demise reached his fellow countrymen, they called a three-day truce in the civil war, and the town of Cabangu changed its name to Santos-Dumont. Combatants on both sides lined up together for miles to file past his open casket in São Paulo. Mourners left flower arrangements in the shapes of dirigibles. The actual funeral was delayed six months, until the war was over and his body could be safely transported to Rio. At the very moment pallbearers lowered his body into the grave that he had prepared for himself, thousands of pilots around the world tipped the wings of their planes in a final gesture of respect.

•

ONE POSTHUMOUS sign of a man's greatness is that his corpse does not make it into the grave in one piece. Not because he donates an organ to science, as Einstein and Lenin did with their brains. But because a fanatical devotee craves a body part to use as a shrine to the great man. Such was Galileo's fate. When the astronomer's body was transferred to a mausoleum in the church of Santa Croce in 1737, nearly a century after his death, one of his disciples, Anton Francesco Gori, broke off the middle finger of his right hand. Today the finger is encased in a small transparent glass casket atop an alabaster column in the Museum of the History of Science in Florence. On the base of the column is a Latin inscription: "This is the finger with which the illustrious hand covered the heavens and indicated their immense space. It pointed to new stars with the marvelous instrument, made of glass, and revealed them to the senses. And thus it was able to reach what Titania could never attain." More than one wag has suggested that Galileo was "giving the finger" to the Church authorities who had tormented him.

Chopin was thinking about Galileo's finger just before his

own death in Paris in 1849. He feared that the Russians who were occupying Poland might stop his body from being re-turned to Warsaw. He told his sister Ludwika that during the autopsy she should remove his heart, conceal it in an urn, and smuggle it back into his homeland. She succeeded, and hid the urn in the catacombs of Warsaw's Holy Cross Church. Although the building was largely destroyed during the 1939 bombing of the Polish capital, the urn survived and today it is part of one of the rebuilt church's columns. Poles flock there to honor Chopin, whose music, at once sad and inspiring, seems to capture the bittersweet history of their country, as does the fact that his heart withstood the Nazis.

WHEN SANTOS-DUMONT died in 1932, Dr. Walther Ha-berfield was in charge of embalming the corpse so that when the war ended, it could be safely transported from São Paulo to Rio for the funeral. Haberfield knew the story of Chopin's burial, and when he was alone in the mortuary, he removed Santos-Dumont's heart. He thought that the organ was unu-sually large, like a bovine's, and he took this as a sign that its owner had been heroically generous and courageous. He bathed the heart in a jar of formaldehyde and took it home with him under his coat. Haberfield did not trust the Vargas government to take proper care of the body. Santos-Dumont belonged to the people, and they should at least have his heart. Twelve years later, in 1944, Haberfield contacted the aero-naut's family and offered the preserved organ. They wanted nothing to do with it, and so he donated it to the government on the condition that it be placed on public display, where anyone could freely visit it and commune with the spirit of "Petite Santos." The doctor's condition was fulfilled—today the custodian of the heart is a small museum at an air-force academy in Campo dos Afonsos, on the outskirts of Rio—but few Brazilians choose to make a pilgrimage there because they do not know of the heart's existence.

IN LATE January 2000, I took an hour cab ride from Copacabana to the museum. I had never seen a human heart before, and I did not know what to expect. The museum's director, a brigadier general who seemed no taller than Santos-Dumont had been, served me bitter coffee in his office under a painting of the aviation pioneer, who looked cheerier and more robust than he did in most of his photos. The general told me that it was an honor to work in a place dedicated to Brazil's most romantic hero. He led me out of his office into a hangar full of old planes. We walked past three or four young soldiers, each a head or two taller than the general. The men sprang to attention and for the first time in my life I was saluted, as I stood there in my blue jeans and sneakers. Not knowing the proper etiquette, I, a child of the sixties, saluted back.

As if it were choreographed, all of them extended their right arms and pointed across the room at a life-size replica of *14-bis*. Made of bamboo and white silk, it was delicately crafted, like a fine piece of Japanese furniture. Even if its canard shape shocked Santos-Dumont's fellow aeronauts, it did not look like the ugly duckling it was sometimes described to be. It was far too elegant and beautiful.

The general left to take a phone call. One of the soldiers who spoke English walked with me to the plane. "Just think," he said, "of the bravery it took to be the first to fly this when no one had flown before him. You or I wouldn't have the courage. Besides, we're too big to fit."

The soldier then led me into a small room full of Santos-Dumont memorabilia. In a glass case was one of his trademark high shirt collars, yellowed with age. There was also a photograph of him at his dandiest. "He is the soul of my country," said the macho-looking soldier. Santos-Dumont, it was evident, had stirred more souls than Tiradentes.

After a long, respectful look at the photo, I said, "May I ask where his heart is?" I tried not to sound too eager.

"It is there," he replied, saluting a ten-inch gold-plated sphere supported by a little winged figure, Icarus probably. The gold was pierced by pinpoint stars that formed the constellations of the Southern Hemisphere. "Inside the sphere," he continued, "is a glass vessel that contains the heart." He saluted the sphere again and stood at attention. I tried peering through the stars to glimpse the heart but they were too tiny, and the light in the room was too poor, to make out much of anything. "The heart is hard to see," he said, "because it is drained of color and floating in preservative, but if you see it, you will have seen the heart of Brazil."

I nodded solemnly and peered again.

"Tell me," he said, relaxing his salute, "why do people in your country insist that the Wright brothers flew first? Nobody saw them on that damn beach. Without witnesses anyone can claim anything. All of Paris watched Santos fly. Why has the world forgotten him? And forgotten his message that the plane should not be used for destruction? How many lives would have been saved?" He paused and looked down at the floor. "If we had heeded his message, there would be no need for a Brazilian Air Force, and I'd be in another line of work." The soldier dabbed his eyes. "It is your mission," he said, "to tell the world about Santos-Dumont. Do it for the glory of Brazil!"

Origins and Acknowledgments

Like most people in the United States, I had never heard of Santos-Dumont before I began this project. In 1996, my friend Matt Freedman returned from a trip to Brazil. Matt knew I was looking for a subject for my next book, and he suggested Santos-Dumont. His presence was felt everywhere in Brazil, Matt told me. People spoke of him reverentially as this larger-than-life figure whose particular amalgam of resolve, inventiveness, showmanship, and generosity represented the spirit of the whole country. I went to Brazil in January 2000 to see for myself.

Before I had even left the United States, I experienced the aeronaut's mystique. In Chicago's O'Hare Airport, I went into a computer boutique to purchase an electrical converter and phone-line adapter so that I could run my laptop in Rio. The sales clerk happened to be Brazilian, and he asked me why I was going to his homeland. I said I was interested in Santos-Dumont. After making sure the other clerk was not watching, he pressed the adapters I needed into my hand. "They're free," he whispered, "for a friend of Santos-Dumont."

Everywhere I went in Brazil I faced a similar reaction. On my first day in Rio, I visited half a dozen used-book stores. My Portuguese was nonexistent, and I had not yet hired a translator, but in each store I said to the man behind the desk, "Santos-Dumont," and the response was always the same. The man nodded approvingly, and the patrons in the store stopped what they were doing, looked at me, and smiled broadly. No bookstore was as generous as the shop in O'Hare, but a few of them cut their prices in half without my asking.

That evening I ate at a *churrascaria,* a traditional barbecue restaurant, near my hotel on the beach. It was one of Rio's famed all-you-can-eat affairs, with a horde of enthusiastic waiters circling the table with sharp knives and long spits of beef, pork, and chicken. On the table there were three stacked coasters: green, yellow, and red. If you put the green one on top, the men with skewers moved in and generously carved meat onto your plate faster than you could eat it. The yellow coaster encouraged them to slow down a bit. And the red one kept them at bay.

I was on the red-coaster stage of the meal when I started flipping through the books on Santos-Dumont that I had acquired earlier in the day. They were heavy on photographs and I lingered over the most intriguing ones. His smashed airship *No. 5* dangling from the roof of the Trocadéro Hotel. Santos-Dumont, dressed in a dark suit, climbing down from *Baladeuse* in front of his apartment on the Champs-Elysées. His sad, deep-set eyes and mustache-covered lips showed no trace of a smile in one picture after another that were taken on presumably happy occasions, after a record-setting flight or the successful test of a new aircraft.

I had not studied the photographs for long when a waiter wielding a skewer of filet mignon approached the table. Had he not seen the red coaster? But he was not there to carve meat. "The Wright brothers. A catapult!" he blurted out in disgust. I tried to talk to him, but that was the extent of his English.

I had been in Rio only a day and already witnessed Santos-Dumont's allure for ordinary Brazilians. I made a decision at dinner to research and write his biography. It was not long before the press in Rio learned of my interest in their illustrious countryman and ran prominent articles about my project. I asked the journalists to mention my E-mail address in their articles along with the request that people contact me who had correspondence, memorabilia, or recollections of the aeronaut. I did this because of the difficulty of tracking down original source material. Because he did not have any direct descendants, his papers are scattered. Moreover, he destroyed his drawings and notebooks, and the sanatoriums he stayed in did not keep his medical records. Some of his airships and other important artifacts in a São Paulo museum had been stolen or vandalized.

The publishing of my E-mail address turned out to be a godsend. Three dozen Brazilians offered to translate material from Portuguese to English, many emphasizing that they would do the work for free, because it was important to them that someone write an honest biography of Santos-Dumont and let the world outside their country know who he was. (The Brazilian biographies of him tended to be short on detail and were fawning and sanitized, and the main English biography, published in 1962, though enormously helpful in getting me started, was incomplete and full of errors, beginning with the assertion that he was the youngest of seven children when, in fact, he was the sixth of eight.) Another four people contacted me whose family heirlooms included eyewitness reminiscences of his Paris flights and correspondence from the man himself. In the end, I received hundreds of letters that he had written; they proved invaluable, particularly in figuring out where the peripatetic aeronaut was at different times in his life, as he moved among his houses in Brazil, residences in Paris and Bénerville, and sanatoriums in Europe.

The memorabilia I received included an unpublished manuscript in which he reflected on his life's work; the last letter he wrote; a crude sketch of the wing of *Demoiselle;* his business card; a bill for his psychiatric treatment; and rare pictures of him as a child. There may be no living witnesses to his flights—you would have to be nearly one hundred to remember them—but I tracked down people who knew him.

In Belo Horizonte, a city two hundred miles west of Rio, I met an eighty-two-year-old waiter at Cantina do Lucas, a late-night wine and pasta joint. As a fourteen-year-old boy, Olympio Peres Munhóz had been the elevator operator at the hotel in Guarujá where Santos-Dumont spent his final days. Munhóz, who was the last person to see him alive, revealed the disturbing circumstances in which he exited the world and the government-ordered doctoring of his death certificate.

In Santos-Dumont's time, his family conspired, along with his friends and the Brazilian military, to suppress any aspects of his life and death that might detract from his status as a hero. Today, though, his relatives seek the truth, and they aided me greatly in my research. In Rio, Sophia Helena Dodsworth Wanderley, Santos-Dumont's grandniece, kindly shared memories of the intrepid aeronaut and showed me the German telescope, now prominent in her living room, that had gotten him in trouble. She was kind enough to let me camp out in her apartment for a week and photocopy the contents of six huge scrapbooks of turn-of-the-century newspaper clippings describing his every move through the air and on the ground. The material belonged to Santos-Dumont himself and surfaced when Sophia's late husband, General Nelson Wanderley, discovered them in a chest in her flooded basement. Santos-Dumont had subscribed to three clipping services, in Paris, London, and New York, which scoured the world's papers for mentions of him. Hundreds of articles were preserved in the scrapbooks, saving me months of research. I am grateful to Sophia and her son, Alberto Dodsworth Wanderley, for their openness, and to Alessandra Blocker at Editora Objetiva, my Brazilian publisher, for delivering a copying machine to Sophia's apartment, after I had failed to persuade countless office-supply stores to do the same.

Another of Santos-Dumont's relatives, Stella Villares Guimarães in São Paulo, shared stories that her grandfather had told about Uncle Alberto. Stella is a graphic designer, and she lugged a large optical scanner from São Paulo to Rio so that we could make copies of the old photographs in Sophia's scrapbooks. It is Stella who is responsible for many of the photos in *Wings of Madness*.

Marcos Villares, a great-grandnephew of Santos-Dumont, informed me that in 1973, on the centennial of the aeronaut's birth, the Brazilian office of Encyclopaedia Britannica held a nationwide contest in which people who had Santos-Dumont mementos were asked to send them to a central clearinghouse. Mario Rangle, who organized the contest, provided me with copies of the hundreds of pages of documents that were submitted.

I also did my own share of combing through newspapers. I spent a year squinting at a microfilm reader as I went through the pages of the *Herald* and other prominent turn-of-the-century periodicals. Most of them were not indexed, and papers like the *Herald* had multiple editions, which meant that I had to look at every page. I turned up another five hundred articles this way. Primary source material came from the Instituto Histórico e Geográfico Brasiliero in Rio, the Biblioteca Nacional in Rio, the Museu Aerospacial in Campos dos Afonsos, the Museu Santos-Dumont in Petrópolis, the Biblioteca Municipal in Santos-Dumont, the Fundação Casa de Cabangu, the British Library, the Royal Aeronautical Society in London, the Royal Society, the Library of Congress, Bibliothèque Forney in Paris, the National Air and Space Museum in Washington, D.C., the Research Libraries of the New York Public Library, the University of Chicago Library, and the Newberry Library in Chicago.

Shante Udon, the chief librarian at Encyclopaedia Britannica in Chicago, helped me track down rare documents. The Missouri Historical Society supplied material on Santos-Dumont's trips to St. Louis. Cartier International opened up its archives on the origin of the wristwatch. The Sem Society in Paris provided details about Santos-Dumont's house in Bénerville.

I employed two researchers, João Marcos Weguelin in Rio and Marina Juliene in Paris, to pursue Portuguese and French sources. Both João and Marina uncovered additional documents and translated them for me. João also accompanied me on trips to Belo Horizonte and Santos-Dumont (Cabangu), and his dedication to this project was invaluable. Here in the States, Sergio Almeida and Eveline Felsten provided additional help with translations.

I also want to thank Sergio Barbosa for providing forty letters that Santos-Dumont had written to his great grandfather, Agenor Barbosa; Henrique Lins de Barros, the director of the Museu de Astronomia e Ciências Afins in Rio for offering initial guidance; Monica Castello Branco Henriques, director of the Fundação Casa de Cabangu, for sharing stories and documents about the aeronaut's childhood home; Tom Crouch at the National Air and Space Museum in Washington, D.C., for reviewing my account of the Wright brothers, and Dan Hagedorn at the same institution for pointing me to Latin American sources; Rebecca Herzig at Bates, Peter Galison at Harvard, and Joseph Corn at Stanford for suggesting background material on the late-nineteenth-century ethos of technological optimism; Will Schwalbe, my editor at Hyperion, and Christopher Potter and Catherine Blyth at Fourth Estate for their cheerful support; Peter Matson for encouraging me when I was lost in a pile of Santos-Dumont clippings; Carolyn Waldron for deft and "squirrelly" copyediting; my brother, Tony, for making me elaborate paper-clip models of *14-bis* and *Demoiselle;* my wife, Ann, for entertaining young Alexander while I slept late in the morning after working long into the night; and lastly Alexander, who gleefully wanted to know if anyone crashes in Daddy's book (he had clearly seen *Those Magnificent Men in Their Flying Machines* one too many times).

Notes

When I quote documents in this book, I preserve the original spelling and punctuation. The two most common sources I cite are the *New York Herald,* abbreviated *NYH,* and *My Air-Ships,* Santos-Dumont's ballooning memoir, abbreviated *SD.* The *Herald*'s sister publication, the *Paris Herald,* was the only English-language newspaper in Paris during Santos-Dumont's flying days, and its publisher, James Gordon Bennett, was his friend as well as an early champion of aeronautics. Consequently, the coverage of Santos-Dumont in both *Herald*s was more extensive than in any other paper. The articles I cite are from the *New York Herald,* but virtually all of them also appeared in the *Paris Herald* on the same day or the day before. Santos-Dumont wrote his memoir at the age of thirty, and so it covers his aeronautical work only through *No. 9, Baladeuse,* and entirely neglects his achievements in heavier-than-air flight. Written in French, the memoir was published in 1904 as *Dans l'air* by Charpentier and Fasquelle in Paris and as *My Air-Ships* by Grant Richards in London and the Century Company in New York (and subsequently released in Portuguese as *Os Meus Balões*). The page numbers I cite are from the Century edition. Original copies are hard to find, but Dover Publications reprinted the memoir in 1973.

Most of the other sources are contemporary newspaper and magazine stories that I found in museum and library clipping files, on microfiche, and in Santos-Dumont's thick scrapbooks. In a few instances, the copies were ripped, which explains incomplete dates or partial titles. One other caveat: Many of the accounts differ in significant details. A fifty-foot flight in one paper may be a five-hundred-foot flight in another. A three-horsepower motor in one article may be a three-hundred-horsepower engine in another. Such differences may be due to printing errors, secondhand reports, ignorant witnesses, or Santos-Dumont's failed and occasionally fabled recollections of his flights. A closer picture of the truth obviously emerges from the examination of multiple sources.

[PROLOGUE]

7   "When the names of those...": [London] *Times,* Nov. 26, 1901.

[CHAPTER 1]

12 "Inhabitants of Europe...": *SD,* p. 18.

13 "I think it is not generally understood...": ibid., pp. 19, 20.

14 "Another endless-chain elevator...": *SD,* pp. 23, 24.

15 "In particular, the moving...": *SD,* pp. 24, 25.

16 "millions of people": *Encyclopaedia Britannica,* vol. 1, ninth edition, 1875, Scribner, p. 189.

17 "model of a dove...": ibid., p. 185.

18 "that his was only a theoretical exercise...": L. T. C. Rolt, *The Aeronauts: A History of Ballooning 1783–1903,* Walker, 1966, p. 25.

19 "No more wonderful scene...": ibid., p. 34.

20 "found that bladders...": *Encyclopaedia Britannica,* p. 189.

20 Information on the Montgolfier brothers and other early balloonists: Rolt, pp. 26–59.

20 "had caused all the old shoes...": M. C. Flammarion, *Travels in the Air,* p. 159, as quoted in Rolt, pp. 28, 29.

22 "The decade of the 1780s . . .": Lee Kennett, *A History of Strategic Bombing,* Scribner, 1982, p. 1.

23 "Pigeon flies!": *SD,* p. 27.

23 "In the long, sun-bathed Brazilian afternoons . . .": ibid., p. 33.

[CHAPTER 2]

26 "All good Americans . . .": *SD,* p. 34.

27 "If we want any work done . . .": Joseph Harris, *The Tallest Tower,* Regnery Gateway, 1979, p. 28.

28 "dizzily ridiculous tower . . .": ibid., p. 28.

28 "at the restaurant on the second platform . . .": ibid., p. 22.

28 "the adventurous wearer . . .": ibid., p. 122.

28 "the harsh smell of gasoline . . .": ibid., p. 144.

29 "Elevators are the exception . . .": Burton Holmes, *Paris,* Chelsea House Publishers, 1998, pp. 90–93.

30 "Polite society proved relatively slow . . .": Eugen Weber, *France, Fin de Siècle,* Harvard University Press, 1986, p. 74. This is a wonderful book on what everyday life was like in Paris at the turn of the century.

30 "the magic, supernatural life": ibid., p. 74.

30 "there were only 30,000 telephones . . .": ibid., p. 76.

30 Discussion of the electric chair: ibid., pp. 73, 74.

32 "You want to make an ascent?": *SD,* pp. 35–37.

34 "I stood there . . .": Santos-Dumont, *O Que eu Vi, O Que nós Veremos,* quoted in Peter Wykeham, *Santos-Dumont,* Harcourt, 1962, pp. 28, 29.

34 "a most dangerous place for a boy . . .": ibid., p. 32.

35 "like the first . . .": *SD,* p. 38.

36 "done at parties, as if one were . . .": Weber, pp. 63, 64. (This practice has been resurrected in the twenty-first century: At chic Long Island garden parties, plastic surgeons administer botox injections after the canapés and before the chardonnay.)

36 "All remarkable women do it . . .": ibid., p. 37.

37 "I regretted bitterly . . .": *SD,* p. 39.

38 "13th July, 12:30 P.M. . . .": Henri Lachambre and Alexis Machuron, *Andrée's Balloon Expedition,* Frederick A. Stokes, 1898, p. 2.

38 "The reading of the book . . .": *SD,* p. 40.

39 "In married life, one has to deal . . .": editors of the Swedish Society of Anthropology and Geography, *Andrée's Story,* Viking, 1930, p. 10.

[CHAPTER 3]

40 "When I asked M. Lachambre how much . . .": *SD,* pp. 40, 41.

41 "Let go, all!": *SD,* pp. 42, 43.

42 "I was frightened": Santos-Dumont, "The Pleasures of Ballooning," *The Independent,* June 1, 1905, p. 1226.

42 "the second great fact . . .": *SD,* p. 44.

43 "making them throw up . . .": ibid., p. 45.

43 "So we had . . .": ibid., p. 46.

44 "Its rubbing . . .": ibid., p. 48.

45 "Observe the treachery . . .": ibid., p. 73.

45 "The lightened balloon . . .": ibid., p. 49.

45 "The Forest of Fontainebleau . . .": ibid., p. 53.

46 "How often have things . . .": ibid., p. 54

47 "It will be too weak": ibid., p. 55.

48 "As I got the pleasure . . .": ibid., p. 59.

48 "I would listen to nothing . . .": ibid., pp. 68–70.

49 "At noon you lunch . . .": ibid., pp. 71, 72.

52 "I was once enamored . . .": Sterling Heilig, "The Dirigible Balloon of M. Santos-Dumont," *The Century Magazine,* November 1901, no. 1, p. 68.

53 "had spent millions . . .": Alberto Santos-Dumont, "How I Became an Aëronaut and My Experience with Air-Ships," Part I, *McClure's Magazine,* vol. XIX, Aug. 1902, p. 314.

53 "happened to be very much perfected . . .": ibid., p. 314.

53 "I looked for the workshop . . .": ibid., p. 314.

54 "I might have had . . .": ibid., p. 314.

55 "From the beginning...": Heilig, *The Century Magazine,* p. 68.

55 "saying that such a thing...": Santos-Dumont, *McClure's,* Part I, p. 315.

56 "a cylinder terminating...": Heilig, *The Century Magazine,* p. 68.

57 "promising good speed...": Santos-Dumont, *McClure's,* Part I, p. 315.

57 "Suppose you are in equilibrium...": Heilig, *The Century Magazine,* p. 69.

59 "Santos, as he prefers...": [Chicago] *Inter Ocean,* "Why I Believe the Airship Is a Commercial Certainty," April 20, 1902.

60 "If you want to commit suicide...": Wykeham, p. 67.

61 "Let go all!": *McClure's,* Part I, p. 316.

62 "Once, on the way to Brazil...": *SD,* pp. 102, 103.

63 "For a while we...": Heilig, *The Century Magazine,* p. 70.

63 "Being inexperienced": *SD,* pp. 93, 94.

64 "As the air-ship grew smaller...": Heilig, *The Century Magazine,* p. 70.

65 "At that moment...": Santos-Dumont, *McClure's,* Part I, p. 316.

65 "For the moment...": *SD,* p. 109.

66 "They were bright young fellows": *SD,* pp. 97, 98.

66 Discussion of Maxim's: H. J. Greenwall, *I'm Going to Maxim's,* Allan Wingate, 1958.

68 "Out to the hangar to get...": ibid., p. 105.

[CHAPTER 4]

69 "a slight accident took place": [London] *Evening News,* 1898.

70 "Finally, after having traversed...": "The Attempted Voyage to Paris," *Aeronautical Journal,* January 1899, p. 19.

70 "An Italian military captive balloon broke loose...": *Aeronautical Journal,* October 1899.

72   "The machinist lay near the motor . . .": *NYH,* "Wife Saw Severo's Balloon Explode," May 13, 1902.

72   "Higher than all . . .": "The Future of American Science," *Science* 1 (Feb. 1883), as quoted in Rebecca Herzig, "In the Name of Science: Suffering, Sacrifice, and the Formation of American Roentgenology," *American Quarterly,* December 2001, pp. 562–581.

73   "cauterization of the bite . . .": Lawrence Altman, *Who Goes First?,* University of California Press, 1987, pp. 23–26, 107–113.

73   Stories about Pasteur and Pettenkofer: ibid., p. 108.

73   "Experimentation permitted on animals": ibid., p. 111.

74   "Even if I had deceived myself": ibid., p. 25.

74   Roentgen's discovery of X rays: Nancy Knight, " 'The New Light' X Rays and Medical Futurism," in Joseph Corn, ed., *Imaging Tomorrow,* MIT Press, 1986, pp. 13–34.

75   "The x-ray mania began early . . .": ibid., p. 14.

76   "The March of Science": *Punch* 110 (1896), p. 117, as cited in ibid., p. 15.

76   Discussion of roentgenology: Herzig, pp. 562–581.

76   "Despite the suffering he has undergone . . .": *New York Times,* "Operated on 72 Times: Roentgenologist Has Lost Eight Fingers and an Eye for Science," March 12, 1926, p. 22, as quoted in ibid., p. 563.

77   "The emerging field of roentgenology . . .": ibid., p. 565.

77   "Scarred and limb-less roentgenologists . . .": ibid., p. 578.

78   "which . . . had all but killed me": Alberto Santos-Dumont, "How I Became an Aëronaut and My Experience with Air-Ships," Part II, *McClure's Magazine,* vol. XIX, Sept. 1902, p. 454.

78   "In those days": *SD,* pp. 127, 128.

79   "This time you . . .": ibid., pp. 128, 129.

79   "The rounder form of this balloon . . .": ibid., p. 30.

80   "He only entered a place . . .": *Jornal do Brasil,* April 25, 1976.

81   "I grow pale": Wykeham, p. 84.

81 "Around that wonderful landmark . . .": Santos-Dumont, *Mc-Clure's*, Part II, p. 454.

81 "Landing in Paris . . .": *SD*, p. 137.

82 "I could well have housed it . . .": ibid., p. 138.

82 "going into air-ship construction . . .": ibid., p. 138.

82 "Even here . . . I had to contend . . .": Heilig, *The Century Magazine,* p. 70.

83 "Although the men had named . . .": *SD,* p. 139.

83 "To fill five hundred . . .": *NYH*, "Steerable Balloon Manoeuvres," Nov. 24, 1899.

83 "between May 1 . . .": *SD*, pp. 156, 157.

84 "balloon was too clumsy . . .": *SD*, p. 139.

85 "friends at the Automobile Club . . .": *New York Times*, "M. Santos Dumont Ready to Test His Balloon," July 10, 1900.

85 "ride on a stick like a witch": *Daily Graphic,* "Aerial Navigation," Oct. 20, 1900.

85 "As Dumont sat straddling . . .": Sterling Heilig, "New Flying Machine," *Washington Star,* June 20, 1900.

86 "It may easily be gathered . . .": Santos-Dumont, *McClure's,* Part II, p. 455.

86 "marvellously ingenious": *Daily Graphic,* "Aerial Navigation," Oct. 20, 1900.

87 "a huge yellow caterpillar": *NYH,* "Aerial Navigation," July 30, 1900.

87 "It was, rather, elliptical . . .": Heilig, *The Century Magazine,* p. 71.

88 "spin round the Bois": *NYH,* "M. Santos-Dumont's Air-Ship Moves Against the Wind," Sept. 20, 1900.

88 "the voyage which M. Dumont . . .": [London] *Daily Express,* "Perilous Ballooning," Sept. 19, 1900.

88 "It would have taken two hours . . .": *NYH,* Sept. 20, 1900.

[CHAPTER 5]

93 Discussion of Chanute and Lancaster drawn from Tom Crouch, *A Dream of Wings,* Norton, 1989, pp. 20–41.

94 "unanimously joined in reviling . . .": ibid., p. 40.

95 "How is it that a turkey-buzzard . . .": Carl Snyder, "The Aerodrome and the Warfare of the Future," *Leslie's Weekly,* July 28, 1896, p. 51.

98 "Its motion was so steady": Ray Coffman, "Prof. Langley First to Make Steady Power Flight Plane," Smithsonian Collection, unidentified clipping.

101 "as one puts a leaf in an extension-table": *SD,* p. 146.

101 "After waiting with the balloon . . .": *SD,* p. 149.

101 "The propeller turned with such force": Santos-Dumont, *McClure's,* Part II, p. 455. This article mistakenly refers to *No. 4* as *No. 5.*

[CHAPTER 6]

102 "Then followed what turned out . . .": *SD,* p. 150.

103 Discussion of customs officials' reaction: Wykeham, p. 108.

104 "Hitherto he had avoided horseless . . .": *NYH,* "President's First Automobile Ride," July 14, 1901.

104 "The carriage swept . . .": *NYH,* "Royal Automobile Upsets the Palace," July 31, 1901.

105 "To drive a spirited horse . . .": *NYH,* "Will Open Park to Automobiles," Nov. 19, 1899.

106 "army corps commanders . . .": *NYH,* "Automobiles for War," Oct. 15, 1900.

106 "Another condition formulated . . .": *SD,* pp. 159–161.

107 "The Santos-Dumont Prize": *SD,* pp. 154, 155.

109 "marvelous, amazing, wonderful . . .": *NYH,* "M. Santos-Dumont Solves the Problem of Aerial Navigation," July 13, 1901.

110 "Paris has been trying to rival New York . . .": *NYH,* "Paris Has a Hot Spell of Its Own, with Many Fatalities," July 14, 1901.

110 Even the privileged could not escape . . . : *NYH,* "Belgium's Queen Overcome by Heat," July 14, 1901.

110 "One man smashed . . .": *NYH,* "Paris Has a Hot Spell of Its Own, with Many Fatalities," July 14, 1901.

111 "The entrance to the park . . .": *NYH,* "Airship Under Control," July 14, 1901.

112 "But for the turn of destiny . . .": [Philadelphia] *American,* "Dumont's Paris Airship Makes a Great Stride in Aeronautics by Sailing Against a Strong Wind," July 15, 1901.

112 "Your evolutions in the air . . .": *SD,* pp. 170, 171.

112 "The hero of the hour . . .": *NYH,* "M. Santos-Dumont Hero of the Hour," July 18, 1901.

113 "a balloon will necessarily always . . .": *Chester* [NY] *Democrat,* "Balloon Navigation Impracticable," July 17, 1901.

113 "The only thing I have accomplished . . .": ibid.

114 "To the Fatherland's Victims": *NYH,* "France Celebrates National Fête," July 15, 1901.

114 "a few planks placed . . .": ibid.

114 "She dined in the Bois . . .": ibid.

115 Comparative death statistics: *NYH,* "Horse Accidents by Far Most Numerous," June 16, 1901.

115 "preying on his mind": *NYH,* "Alienist Doctor Goes Mad," July 21, 1901.

116 The heat wave had continued . . . : *NYH,* "Many Persons in Europe Killed by Lightning," July 22, 1901.

117 The inclement weather did not stop . . . : *NYH,* "Parisians Out to See Airship," July 22, 1901.

117 "The flying machine was almost . . .": W. L. McAlpin, "Santos Dumont and His Air Ship," *Munsey's Magazine,* [month unknown], 1902.

118 "including a large number . . .": *NYH,* "Santos-Dumont Tries Again," July 30, 1901.

118 A few days later . . . : Santos-Dumont: *New York Times,* "Dirigible Balloon Fails," Aug. 5, 1901.

118 "M. Santos-Dumont is nothing better . . .": *NYH,* "Like Another Dreyfus Affair," Aug. 2, 1902.

120 "another Dreyfus affair": ibid.

120 "It is a very kindly idea . . .": *NYH,* "Parisians Out to See Airship," July 22, 1901.

120 The Brazilian government soon came through . . . : *NYH,* "Applause from Brazil," Aug. 14, 1901.

120 "August 1, 1901. Monsieur Santos-Dumont . . .": *SD,* p. 173.

121 "We saw the balloon . . .": [London] *Daily Express,* Aug. 9, 1901.

122 "But here I was competing . . .": *SD,* p. 177.

122 "Two thousand feet in the air . . .": [London] *Daily Express,* Aug. 9, 1901.

123 "It may have seemed a terrific fall . . .": *SD,* p. 178.

124 "One could hear screams . . .": [London] *Daily Express,* Aug. 9, 1901.

124 "The whole of the keel . . .": *SD,* p. 181.

124 "The keel, in spite of my weight . . .": ibid., p. 182.

125 "The operation was painful . . .": ibid., p. 185.

125 "The reception of M. Santos-Dumont . . .": *NYH,* "Santos-Dumont's Escape," Aug. 9, 1901.

125 "He was so affected by the danger . . .": ibid.

126 "M. Santos-Dumont looked . . .": *Daily Telegraph,* Aug. 9, 1901.

126 "I am afraid the experiments . . .": *NYH,* "Santos-Dumont's Escape," Aug. 9, 1901.

[C H A P T E R 7]

127 The framework, which had surprisingly withstood . . . : *NYH,* "M. Santos-Dumont Plans New Airship," Aug. 10, 1901.

127 Like the price of absinthe . . . : *NYH,* "Price of Absinthe Raised in Paris," Aug. 18, 1901.

128 Like the new spittoons . . . : *NYH,* "Dr. Koch's Theory Discredited," Aug. 18, 1901.

128 Like the infighting . . . : *NYH,* "Parasols for Horses," Aug. 18, 1901.

128 A laborer named Simon . . . : *NYH,* "Four Days in Well and Found Alive," Aug. 11, 1901.

128 Organizers of the Pan American Exposition . . . : *Buffalo Courier,* "M. Dumont's Airship Expected," Aug. 23, 1901.

128 "I confess . . . that the idea of breaking all records . . .": *New York Journal,* "Around the World in an Airship," Oct. 13, 1901.

128 "It appears that one of the greatest troubles . . .": *NYH,* "Actresses Beset Paris Aeronauts," Sept. 3, 1901.

129 At the end of August . . . : *NYH,* "M. Santos-Dumont Sued for Damages to Tiled Roof," Aug. 27, 1901.

129 "the vexations of who knows what . . .": *SD,* p. 225.

130 "Monsieur le President . . .": *NYH,* "M. Santos-Dumont Makes a Protest," Sept. 11, 1901.

134 "The frame bearing the weight of the motor . . .": *NYH,* Sept. 21, 1901.

134 "In aeronautics . . . especially is verified . . .": *NYH,* "How Airship Was Wrecked," Sept. 23, 1901.

135 "Such accidents I have always taken philosophically . . .": *SD,* p. 201.

135 "M. Santos Dumont is to be pitied": *Rangoon Gazette,* "Ballooning," Oct. 9, 1901.

136 "with perfect docility": *NYH,* "M. Santos-Dumont Successful," Oct. 11, 1901.

137 "wheeled around in circles": ibid.

138 "When the airship reached a point . . .": [Boston] *Post,* "Santos-Dumont Describes His Journey through the Air on Saturday," Oct. 21, 1901.

138 "a wild stampede of foot-passengers . . .": *NYH,* "M. Santos-Dumont Rounds the Eiffel Tower," Oct. 20, 1901.

139 "Here was another conqueror . . .": *Westminster Gazette,* "The Great Airship Triumph," Oct. 21, 1901.

139 "The rest of the trip was remarkably intoxicating": [Boston] *Post,* Oct. 21, 1901.

140 "Have I won the prize?": *Philadelphia Inquirer,* "Santos-Dumont King of the Air," Oct. 20, 1901.

140 "My friend, . . . you have lost the prize by forty seconds": *Daily Messenger,* "The Santos-Dumont Balloon," Oct. 20, 1901.

141 "For my part, I consider that you have won the prize!": *Philadelphia Inquirer,* Oct. 20, 1901.

141 "I could have landed": [Boston] *Post,* Oct. 21, 1901.

141 "moral victory": *NYH,* "M. Santos-Dumont Rounds the Eiffel Tower," Oct. 20, 1901.

142 "When M. Santos-Dumont's well-known little electric automobile . . .": ibid.

142 "So far as I am concerned": *Philadelphia Inquirer,* Oct. 20, 1901.

142 "another Dreyfus affair": *NYH,* "Public Favors M. Santos-Dumont," Oct. 21, 1901.

143 "Santos, the great name of the week . . .": *NYH,* "Ballooning: Mr. Santos-Dumont About to Be Immortalized by the Tailors and Toymakers," Oct. 15, 1901.

143 "The latest fashion from Paris in lady's hats . . .": *Dry Goods Economist,* New York, Dec. 21, 1901.

143 "Even the tiniest toddlers . . .": *NYH,* "M. Santos-Dumont Very Popular," Nov. 7, 1901.

143 "It is a sign of the times that the toy balloons . . .": *NYH,* Oct. 28, 1901.

144 "The fact that it cannot fly . . .": *Denver Times,* "Toy Flying Machines," Jan. 6, 1902.

145 "the press had made Santos-Dumont's face . . .": *La Vélo,* Nov. 9, 1901.

145 "The Parisian populace must always have a hero . . .": *Dispatch,* "Paris Idolatry Now Rests Upon Hero of Airship," Nov. 6, 1901.

146 "aerial sole, very delicate . . .": *Daily Telegraph,* "A Glória de Santos-Dumont," Nov. 11, 1901.

147 "When M. Santos-Dumont stood up to respond": *Daily Messenger,* "Santos-Dumont in London," Nov. 26, 1901.

147 "Great British nation . . .": *Sketch,* "The Aerial Navigator," Nov. 12, 1902.

147 "If one could create an aeronaut . . .": [London] *Daily News,* "M. Santos-Dumont in London," Nov. 23, 1901.

148 "Santos-Dumont is the very last person in the world . . .": *Brighton Standard,* Jan. 4, 1902.

148 "Yes . . . I have had many accidents . . .": [London] *Daily News,* Nov. 23, 1901.

148 "Yes . . . I also come to London . . .": ibid.

[CHAPTER 8]

150 "It will no longer be an absurdity . . .": *Westminster Gazette,* "The Great Airship Triumph," Oct. 21, 1901.

151 "was not sufficiently . . .": James J. Horgan, *City of Flight: The History of Aviation in St. Louis,* The Patrice Press, Gerald, Missouri, 1984, p. 44.

152 "change the whole conditions of warfare . . .": Maj. Charles B. van Pelt, "The Aerodrome That Almost Flew," *American History Illustrated,* Dec. 1966, p. 46

152 "will make armies a jest . . .": Carl Snyder, "The Aerodrome and the Warfare of the Future," *Leslie's Weekly,* July 28, 1896, p. 55.

153 "In all great wars hitherto . . .": ibid.

154 Details of Gatling's work: John Ellis, *The Social History of the Machine Gun,* Croom Helm, 1975.

154 "It may be interesting to you to know . . .": ibid., p. 26.

155 "The bulk of these officers . . .": ibid., p. 16.

155 "aeroplanes and tanks . . .": ibid., p. 17.

155 "Without the handful of machine guns": ibid., p. 18.

156 Discussion of Alfred Nobel: Nicholas Halasz, *Nobel: A Biography,* Robert Hale Limited, 1960.

156 "The abolition of slavery . . .": ibid., p. 154.

156 "Perhaps my factories will . . .": ibid., pp. 158, 159.

157 "A man who would discover . . .": ibid., p. 159.

157 "If Andrée reaches his goal . . .": ibid., p. 173.

158 "It is my express will . . .": ibid., pp. 180, 183, 184.

159 "The twentieth century . . .": ibid., p. 185.

159 "in vain for the city . . .": Luis Alvarez, *Alvarez: Adventures of a Physicist,* Basic Books, 1987, p. 7.

160 "The story of our mission . . .": ibid., p. 8.

[C H A P T E R 9]

161 "Do you remember the time, my dear Alberto . . .": *SD,* pp. 28, 29.

162 "education as an airship captain": ibid., p. 226.

162 "Suppose you buy a new bicycle . . .": ibid., pp. 218–221.

162 "The winning of the Deutsch Prize": ibid., p. 229.

163 "And of the twenty-six . . .": ibid., p. 222.

163 "a number of American 'millionaire' owners . . .": *NYH,* "M. Santos-Dumont on Mediterranean," Nov. 3, 1901.

165 "performed at the end of its guide-rope . . .": *SD,* p. 236.

165 "It had to be solidly constructed": ibid., p. 233.

166 "Their equilibrium was so well calculated . . .": ibid., p. 234.

166 "the blue Mediterranean into the Red Sea": *NYH,* "M. Santos-Dumont's Flight Checked. Riviera Official Thought He Was Turning Blue Mediterranean into Red Sea," Jan. 26, 1901.

167 "It will be my most ambitious effort . . .": [New York] *Journal,* "Hey, for a Flight to Africa! Is Santos-Dumont's Cry To-Day," Jan. 26, 1902.

167 "a miscalculation had been made . . .": *SD,* p. 241.

168 "They were at last able to catch . . .": ibid., p. 242.

168 "Straight as a dart the air-ship sped . . .": ibid., p. 245.

169 "as a needle is threaded . . .": ibid., p. 246.

169 "I will not ask you to do so much . . .": ibid., p. 246.

169 "I think it will be possible to cross the Atlantic . . .": Santos-

Dumont, *Baltimore American,* "Travel by Balloon," Jan. 5, 1902.

170 "The other day M. Henri Rochefort . . .": *Daily Express,* "Remarkable Meeting," Feb. 8, 1902.

171 *"Por Mares Nunca D'antes Navegados!"*: *NYH,* "M. Santos-Dumont Out for a Fly," Feb. 11, 1902.

171 "The guide-rope held me at a steady altitude . . .": *SD,* pp. 248–252.

171 "Their sails were full-bellied . . .": ibid., pp. 252, 253.

172 "Those with the prince": ibid., p. 263.

173 "It was a glorious day": [London] *Daily Mail,* "Airship Wrecked," Feb. 15, 1902.

175 "As a consequence, the hydrogen nearest the silk . . .": *SD,* pp. 284, 285.

175 "Looking back over all my varied experiments": ibid., p. 293.

176 "The intrepid aeronaut has decided to decline . . .": [London] *Times,* Feb. 23, 1902.

[C H A P T E R   1 0]

178 "Mr. Dumont has recovered from his immersion . . .": *Daily Chronicle,* "M. Santos-Dumont Moves His Headquarters to London," March 5, 1902.

178 "It was a rather unusual way . . .": *NYH,* "Aeronaut's Farewell," March 5, 1902.

179 "My new *No.* 7 has forty-five horsepower . . .": *Philadelphia Record,* "Dumont Longs for America," March 9, 1902.

180 "If I were to choose my nationality": ibid.

180 "to go up the East River . . .": *NYH,* "To Fly Over the Brooklyn Bridge," March 5, 1902.

180 "With the English I am at home": *Philadelphia Record,* "Dumont Longs for America," March 9, 1902.

180 "bring competition. And I like . . .": *Daily Express,* "Dumont Wants Rivals," March 6, 1902.

181 "He was driven from Victoria Station . . .": *Senhor Santos Du-*

mont's Reception in London, 1901: The Aero Club Banquet, private minutes, British Library.

181 "The suggestion that there might be difficulty . . .": *NYH,* "London to Have Ambulances," March 9, 1902.

181 "things which might have shocked . . .": [Pittsburgh] *Dispatch,* "Santos-Dumont Never Heard of Tariff," April 12, 1902.

182 "The man who flies in the air smiled . . .": ibid.

182 "the great air port of the New World": newspaper unknown.

182 "as calmly as a farmer . . .": newspaper unknown.

183 "His eyes are reddish hazel . . .": *Pittsburgh Press,* "Santos-Dumont Forecasts Days of Aerial Navigation," April 12, 1902.

183 "M. Santos-Dumont, the 'King' . . .": *New-York Mail and Express,* "Santos-Dumont Knits and Sews," April 19, 1902.

186 "Pike's Peak the Future Summer Home of the Wealthy": *New York Journal,* April 14, 1902.

186 "Perhaps Santos-Dumont Will Solve the Bridge Problem": *Brooklyn Daily Eagle,* March 5, 1902.

186 the chest of a zaftig woman: *Brooklyn Daily Eagle,* May 18, 1902.

186 "These are much taller": newspaper unknown.

187 "I was down in Florida . . .": *Philadelphia Telegraph,* "Tom Edison's Airship Talk with Santos-Dumont," May 2, 1902. The entire conversation between Edison and Santos-Dumont comes from this article.

191 "I am happy to see you . . .": *NYH,* "President Would Take Trip in Air," April 17, 1902.

191 "The Rough Riders of the Future": *Brooklyn Eagle,* April 17, 1902.

192 "Then you will take me up . . .": *NYH,* "President Would Take Trip in Air," April 17, 1902.

192 "Until some absolutely reliable motor . . .": *Pall Mall Gazette,* "M. Santos-Dumont," March 4, 1902.

194 "There are plenty of men . . .": *New York Journal,* "Edison Would Join Aerial Club," April 14, 1902.

194 "Bowed by the weight of four score years . . .": *New York Journal,* "Airship Is Useless, Says Lord Kelvin," April 20, 1902.

195 "Lord Kelvin's statement that my airship . . .": *New York Times,* "Santos-Dumont Sails Away," May 2, 1902.

196 "I am an amateur": *NYH,* "Amateur Aeronautics," May [day unknown], 1902.

196 "I shall leave my *No.* 7 in this country": *NYH,* "Looks to America to Perfect Airship," May 1, 1902.

196 "If any one will give me a million dollars . . .": *New York Times,* May 2, 1902.

196 "Certainly I meant it": [Philadelphia] *Telegraph,* "Can Build an Airship to Cross the Ocean," May [day unknown], 1902.

199 "It is an outrage . . .": [London] *Sun,* "Santos-Dumont's Loss," May 28, 1901.

199 "They say we aeronauts are all crazy": [Philadelphia] *Evening Standard,* "Balloon Cut into Ribbons," May 28, 1902.

199 "This cannot be": [London] *Daily Express,* "Airship Mystery," May 28, 1902.

199 "by the action of the gas . . .": [London] *Morning Leader,* May 21, 1902.

200 "When we first noticed the damage": [London] *Daily Express,* "Airship Mystery," May 28, 1902.

200 "a score of private watchmen . . .": [Brooklyn] *Standard Herald,* "Santos-Dumont's Airship Will Be Tested at Brighton Beach," July 12, 1902.

200 "I don't intend to prepare my machine . . .": *NYH,* "Will Fly Only for Definite Object," July 5, 1902.

201 "Am sailing 17th . . .": *Brooklyn Daily Eagle,* "Santos-Dumont Is Coming," July 13, 1902.

201 "delicate workmanship": *NYH,* "Santos-Dumont's Air Ship Inflated, Ready to Fly When Owner Arrives," July 20, 1902.

202 "as a horseman would inspect a horse . . .": *NYH,* "M. Santos-Dumont Inspects His Air Ship *No. 6,*" July 24, 1902.

202 "Newport, R.I., Thursday . . .": *NYH,* "M. Santos-Dumont Flies to Rescue," Aug. 1 (or 11?), 1902.

203 "Two hundred people were badly frightened": *NYH,* "Airship Propeller Frightens Crowd," Aug. 11, 1901.

203 "a loud report followed by a tremendous crash . . .": [Rochester] *Herald,* "Dumont's Airship Damaged," Aug. 12, 1902.

204 "I am disgusted with it all": *NYH,* "Santos-Dumont Hurriedly Sails," Aug. 15, 1902.

205 "It was only yesterday . . .": [Lafayette] *Mall,* Aug. 16, 1902.

205 "the greatest disappointment of my life": *NYH,* "M. Santos-Dumont Is Disappointed," Aug. 26, 1902.

[C H A P T E R 1 1]

208 "the smallest of possible dirigibles": *SD,* p. 313.

209 "sought to prove to an unbelieving world . . .": Helen Waterhouse, "La premiere aero-chauffeuse," *Sportsman Pilot,* July 1933.

209 "I had just sat down . . .": *L'Illustration,* July 4, 1903.

210 "Knowing that the feat . . .": *SD,* pp. 319–327.

212 "The boy will surely make an airship captain . . .": ibid., p. 327.

212 "confessed an extraordinary desire . . .": ibid., pp. 328–331.

213 "I've flown dirigibles myself": Waterhouse.

214 "He showed me how to steer the big rudder . . .": [Washington] *Sunday Star,* "The First Woman to Fly a Dirigible," June 25, 1933.

215 "Santos! Santos!": *Milwaukee Journal,* "Society Girl Flew Before the Wrights," Aug. 20, 1933.

215 "I will never forget . . .": *Christian Science Monitor,* "Only Woman to Fly Dirigible Eligible for Early Bird Honor," July 10, 1933.

216 It was also her fetching, if bulky: Waterhouse.

216 "In my long black and white foulard . . .": *Milwaukee Journal,* "Society Girl Flew Before the Wrights," Aug. 20, 1933.

216 *"la premiere aero-chauffeuse du monde"*: *Christian Science Monitor,* "Only Woman to Fly Dirigible Eligible for Early Bird Honor," July 10, 1933.

216 *"C'est fou!"*: ibid.

220 "the life of a cannon-ball . . .": William Sanson, *Proust and His World,* Charles Scribner's Sons, 1973, p. 75.

220 "The airplanes which a few hours . . .": *Remembrance of Things Past* as quoted in Stephen Kern, *The Culture of Time and Space 1880–1918,* Harvard University Press, 1983, p. 245.

221 "He was quite odd": Amália Dumont, "Reminiscence," *O Globo.*

221 "Boston, Mass., Thursday . . .": *NYH,* "Santos-Dumont Named in a Divorce Suit," Jan. 16, 1903.

222 "There is not the slightest . . .": *NYH,* Jan. 18, 1903.

223 "He had no special lady . . .": Author's conversation with Sophia Helena Dodsworth Wanderley in June 2224.

224 "was deeply distressed when the tongue of malicious . . .": Henrique Dumont Villares, *Santos-Dumont "The Father of Aviation,"* [no publisher cited], 1956, p. 28.

227 "a handful of mortar": Walter T. Bonney, "Prelude to Kitty Hawk Part IV," *Pegasus,* Aug. 1953, p. 12.

227 "An Ingenious Man . . .": John M. Taylor, "The Man Who Didn't Invent the Airplane," *Yankee,* Nov. 1981, p. 223.

228 "I see that Langley . . .": Stephen Kirk, *First in Flight: The Wright Brothers in North Carolina,* John F. Blair, 1995, p. 174.

229 "Langley's Dream Develops . . .": ibid., p. 192.

229 "Perhaps if Professor Langley . . .": Bonney, p. 14.

229 "The professor does not . . .": Kirk, p. 192.

229 "In the past . . . we have paid our respects . . .": Kirk, p. 193.

229 "Failure in the Aerodrome itself . . .": John Tierney, "Langley's Aerodrop," *Science '82,* March 1982, p. 82.

229 "Tell Langley": ibid., p. 82.

229 "If it is going to cost . . .": *American History Illustrated,* [date unknown], p. 53.

230 "The newspapers report the death . . .": Taylor, p. 224.

231 "the first man-carrying aeroplane . . .": ibid., p. 227.

231 "I believe that my course . . .": Tierney, p. 82.

232 "The two metal tubes . . .": ibid., p. 83.

233 "Flying Machine Soars Three Miles . . .": Kirk, p. 190.

233 "Dayton Boys Emulate Great Santos-Dumont": ibid., p. 193.

[CHAPTER 12]

234 "the chief interest which the tests . . .": *New York Times,* "Air Sailing," Jan. 14, 1904.

235 Information on John Wise: James Horgan, *City of Flight: The History of Aviation in St. Louis,* pp. 42–53.

235 "I had never seen . . .": ibid., p. 42.

236 "a large quantity of cold chickens . . .": ibid., p. 46.

237 "If only one man can go . . .": ibid., p. 52.

238 "I expect that at least 150 airships . . .": James Horgan, "Aeronautics at the World's Fair of 1904," *Bulletin,* Missouri Historical Society, April 1968.

238 "The speed expected is too great": "A Letter from Leo Stevens," *Scientific American,* March 26, 1904.

238 "More than the most ardent automobilist . . .": *NYH,* "M. Santos-Dumont Is Confident of Winning Prize Airship Race," January 1904.

240 "They even accuse me of growing stout": *NYH,* "Santos Dumont to Enter Contest," March 3, 1904.

240 "Lost sixty horse power engine . . .": James Horgan, "The Strange Death of Santos-Dumont *Number 7,*" *AAHS Journal,* Sept. 1968.

241 "I have never raced this airship": *New York Times,* "Santos-Dumont Here to Fly for Airship Prize," June 18, 1904.

242 "To me the cutting was done . . .": *New York Times,* "Dumont's Big Airship Slashed by a Vandal," June 29, 1904.

242 "It is an outrage! . . .": *NYH,* "Santos-Dumont Airship Slashed," June 29, 1904.

243 "The bag of the airship . . .": *NYH,* "M. Dumont Orders New Airship Bag," June 30, 1904.

244 "so wrought up . . .": *NYH,* June 29, 1904.

244 "If Professor Meyers repairs the bag . . .": *New York Times,* "Accuses Santos-Dumont," June 30, 1904.

244 "I learn from Lieut. Walsh . . .": ibid.

245 "How is it conceivable that I destroyed . . .": *NYH,* " 'I Cut It? Absurd!' M. Santos-Dumont," July 1, 1904.

250 "jumped at the Russian offer of $200,000 . . .": *St. Louis Post-Dispatch,* "Russia Figures in Cutting of Airship Here," Oct. 22, 1907.

[CHAPTER 13]

251 "As a youth": *SD,* p. 353.

252 "I have never sat down to work seriously . . .": *Lecture pour Tous,* Jan. 1, 1914.

254 "And if I were to tell you . . .": *Je sais tout,* Feb. 15, 1905.

256 "All attempts at artificial . . .": [London] *Times,* [day unknown], 1905.

256 "every one present saw the wheels . . .": *NYH,* "Aeroplane Raised by Small Motor," Aug. 23, 1906.

257 "But he flew": *NYH,* "Santos-Dumont Flies 37 Feet," Sept. 14, 1906.

258 "1—We have just come down in a balloon . . .": *NYH,* "Aeronauts of Seven Nations Contest for International Cup," Oct. 1, 1906.

258 "bottles of mulligatawny . . .": London *Tribune* quoted in *NYH,* "Aero Club Busy on Balloon Race," Sept. 29, 1906.

259 "the great balloon race . . .": *Pelican* quoted in *NYH,* Sept. 29, 1906.

259 "The crowd was stirred . . .": *NYH,* "Santos-Dumont Wins $10,000 Aerial Prize," Oct. 24, 1906.

260 "I really do not know why . . .": ibid.

260 "He raised the head . . .": ibid.

263 "does not appeal to us . . .": *NYH,* "Dayton Aeronauts Are Not Surprised," Nov. 13, 1906.

263 "If he has gone more than 300 feet . . .": Crouch, *The Bishop's Boys,* p. 317.

263 "M. Santos-Dumont in a few months . . .": *NYH,* "Santos-Dumont Aeroplane Simple," Oct. 25, 1906.

264 "Between the time of Santos's short hop . . .": Crouch, *The Bishop's Boys,* p. 317.

267 "the Wrights proved to be . . .": ibid., p. 301.

267 "warped by their desire for great wealth": ibid., pp. 301, 302.

267 "Is it possible . . .": *Scientific American,* "The Wright Aeroplane and Its Fabled Performances," Jan. 13, 1906, p. 40.

267 "fliers or liars": *NYH,* "Fliers or Liars," Feb. 10, 1906.

269 "The only birds that talk . . .": Nancy Winters, *Man Flies,* The Ecco Press, 1997, p. 128.

270 "Princes and millionaires": Crouch, *The Bishop's Boys,* pp. 382, 383.

270 "The smallest details of their lives . . .": ibid., p. 387.

271 "It was, I may say now": *L'homme Mecanique,* 1929, unpublished manuscript, from the collection of General Nelson Wanderley.

271 "virtually in his lap, his legs straddling . . .": Henry P. Palmer Jr., "The Birdcage Parasol," *Flying,* Oct. 1960.

272 "became so casual about his aerial outings . . .": John Underwood, "The Gift of Alberto Santos-Dumont," source unknown.

273 "This altitude was, as a matter of fact . . .": Henrique Lins de Barros, *Alberto Santos-Dumont,* Editora Index, 1986, pp. 115–118. A marvelous bilingual work (Portuguese and English) filled with great photographs and illustrations of Santos-Dumont and his work.

274 "in a few months . . . home-built *Demoiselles* were hopping . . .": Underwood.

274 "Apart from bumps and bruises . . .": ibid.

274 "a bracing wire snapped, collapsing a wing . . .": ibid.

[CHAPTER 14]

279 "all advances made by modern science . . .": Michael Adas, *Machines as the Measure of Man,* Cornell University Press, 1989, p. 366.

279 "I use a knife to slice gruyere": unidentified newspaper.

280 "infantry who had . . .": ibid., p. 235.

280 "The famed Krupp works . . .": Adas, p. 367.

280 "a single regiment of field guns . . .": ibid., p. 367.

280 "a war of engineers and chemists . . .": David Wragg, *The Offensive Weapon,* Robert Hale, 1986, p. 1.

280 "Railroads made it possible . . .": Adas, p. 367.

281 Discussion of the Hague conferences: Barbara Tuchman, *The Proud Tower,* Macmillan, 1966, pp. 229–288.

281 "We belong to each other": Tuchman, p. 240.

281 "If your Emperor commands you to do so": ibid., p. 240.

282 "The civilized soldier when shot . . .": ibid., p. 262.

284 "This transformation of geography . . .": Wykeham, p. 234.

284 "I only followed . . .": Wykeham, p. 234.

285 "Why not a hundred years earlier?": Curtis Prendergast, *The First Aviators,* Time-Life Books, 1981, p. 49.

285 "only ten men in the world . . .": Crouch, *The Bishop's Boys,* p. 404.

286 "Love letters . . .": Roger Bilstein, *Flight in America,* The Johns Hopkins University Press, 1984, p. 17.

286 "The idea of a man flying up . . .": ibid., p. 26.

288 "Thirty thousand eyes . . .": Joseph Corn, *The Winged Gospel,* Oxford University Press, 1983, p. 4.

288 "Never have I seen . . .": ibid., p. 4.

289 "Beachey challenged the concrete walls . . .": Bilstein, pp. 20, 21.

290 "I flew tight circles . . .": unidentified newspaper.

290 "The crowd . . . gaped at the wonders . . .": Bilstein p. 25.

291 "There was an old woman . . .": ibid., p. 25.

[C H A P T E R 15]

294 "sent shards of metal . . .": Adas, pp. 370, 371.

295 "It is a youth-intoxicated profession . . .": Edgar Middletown, *Glorious Exploits of the Air,* D. Appleton & Company, 1918, pp. 14, 15.

295 "The 'bon camaraderie' that has developed . . .": ibid., pp. 189, 190.

296 "military honours in unstinting . . .": Floyd Gibbons, *The Red Knight of Germany,* The Sun Dial Press, 1927, p. 2.

296 "He fought, not with hate . . .": ibid., p. 2.

296 "Some went down like flaming comets . . .": ibid., p. 3.

296 Casualty statistics for airmen: John H. Morrow Jr., *The Great War in the Air,* Smithsonian Institution Press, 1993, p. 367. This book is the definitive work on the subject.

296 "They skim like armed swallows . . .": ibid., p. 365.

298 "the first of some five hundred Parisians . . .": Lee Kennett, *The History of Strategic Bombing,* p. 20.

299 "In grim statistical terms": ibid., p. 25.

300 "The Aeroplane . . . has made war so terrible . . .": Bilstein, p. 39.

[C H A P T E R 16]

302 "He now believes that he is more infamous . . .": Wykeham, p. 247.

303 "Those, who like myself were the humble pioneers . . .": Henrique Dumont Villares, pp. 43, 44.

304 "Dear friend, I couldn't sleep . . .": excerpts from the diary of A. Camillo de Oliveira, 254 from the collection of General Nelson Wanderley.

304 "We know you have good relations with Mr. Santos": ibid.

305 "I have always requested . . .": Barros, p. 131.

306 Discussion of Ford and aerial cars: Corn, *The Winged Gospel,* pp. 91–111.

307 "A New York . . .": ibid., p. 95.

308 "The partisans of the Wright brothers assert . . .": *L'homme Mecanique,* from the collection of General Nelson Wanderley.

308 "had automatic legs on all of their machines . . .": ibid.

310 "I never thought that my invention . . .": Author's conversation with Olympio Peres Munhóz in June 2000.

What Santos-Dumont Wrote

A Conquista do Ar Spelo Aeronauta Brasiliero Santos-Dumont ("Brazilian Aeronaut Santos-Dumont's Conquest of the Air"), 1901. A slim pamphlet not available in English.

"Travel by Balloon," *Baltimore American,* Jan. 5, 1902.

"How I Became an Aëronaut and My Experience with Air-Ships," Part I, *McClure's Magazine,* vol. XIX, Aug. 1902.

"How I Became an Aëronaut and My Experience with Air-Ships," Part II, *McClure's Magazine,* vol. XIX, Sept. 1902.

My Air-Ships, The Century Company, New York, 1904.

"The Sensations and Emotions of Aerial Navigation," *The Pall Mall Magazine,* 1904.

"Ce Que Je Ferai, Ce Que L'on Fera" ("What I Will Do, What They Will Do"), *Je sais tout*, Feb. 15, 1905.

"The Pleasures of Ballooning," *The Independent,* June 1, 1905.

O Que eu Vi, O Que nós Veremos ("What I Saw, What We Shall See"), 1918 (not available in English translation).

L'homme Mecanique ("The Mechanical Man"), 1929 (unpublished French manuscript).

What Santos-Dumont Read

Octave Chanute, *Progress in Flying Machines,* New York, 1894.

Victor Hugo, *Les misérables* (found in his house in Petrópolis after his death).

Henri Lachambre and Alexis Machuron, *Andrée's Balloon Expedition,* Frederick A. Stokes, 1898.

Adolfo Venturi, *Botticelli,* A. Zwemmer, 1927 (one of the books he bound by hand while he was in a Swiss sanatorium).

Jules Verne, *Five Weeks in a Balloon,* 1863.

——, *A Journey to the Center of the Earth,* 1864.

——, *From the Earth to the Moon,* 1866.

——, *Twenty Thousand Leagues Under the Sea,* 1870.

——, *Around the World in Eighty Days,* 1873.

——, *Mysterious Island,* 1874.

——, *Master of the World,* 1904.

H. G. Wells, *The War of the Worlds,* 1898.

——, *The War in the Air,* 1908.

What Santos-Dumont Made

1883 miniature hot-air balloons from tissue paper

1883 rubber-band-powered toy wooden plane

1897 first balloon flight (with Alexis Machuron)

1898 *Brazil,* hydrogen balloon
    appearance: small pear-shaped gasbag with long rigging
    dimensions: 20 feet in diameter
    gas capacity: 4,000 cubic feet
    characteristics: lightweight Japanese silk
    performance: more than 200 ascensions

1898 *No. 1,* one-person airship
    appearance: a cylinder terminating in cones
    dimensions: 82.5 feet long, 11.5 feet at greatest diameter
    gas capacity: 6,454 cubic feet
    motor: 3.5-horsepower converted tricycle engine
    characteristics: engine attached to basket; no chemise; movable
      weights to shift center of gravity; air pump; silk rudder;
      guide rope; pusher propeller

performance: smashed into trees (Sept. 18); collapsed under atmospheric pressure (Sept. 20)

1899 *No. 2,* one-person airship
   appearance: similar to *No. 1*
   dimensions: 82.5 feet
   gas capacity: 7,062 cubic feet
   motor: 3.5 horsepower
   characteristics: small rotary fan to assist air pump in filling internal balloonet; pusher propeller
   performance: collapsed before ascending above trees (May 11)

1899 *No. 3,* one-person airship
   appearance: pudgy football
   dimensions: 66 feet long, 25 feet at greatest diameter
   gas capacity: 17,650 cubic feet
   motor: 3.5 horsepower
   characteristics: no balloonet or air pump; lamp gas; 33-foot bamboo pole to provide rigidity; pusher propeller
   performance: attained 12 mph; 20-minute flight around Eiffel Tower (Nov. 13); dozens of trips; record 23 hours aloft

1900 *No. 4,* one-person airship
   appearance: elliptical; "huge yellow caterpillar"
   dimensions: 95 feet long, 17 feet at greatest diameter
   gas capacity: 14,800 cubic feet
   motor: 7 horsepower
   characteristics: no balloon basket; aluminum ventilator; exposed bicycle seat; first puller propeller; huge hexagonal silk rudder
   performance: tethered ascent in stormy weather before the International Aeronautical Congress (Sept. 19)

1900 reconstructed *No. 4,* one-person airship
   appearance: elliptical
   dimensions: 108 feet long

gas capacity: unknown

characteristics: added silk to the balloon envelope "as one puts a leaf in an extension-table"

performance: unstable; never tested

1901 *No. 5,* one-person airship

appearance: elliptical

dimensions: 118 feet

gas capacity: 19,420 cubic feet

characteristics: 60-foot keel braced by piano wire; pusher propeller; taxed by customs officials as a piece of fine cabinetry

performance: crashed into the Rothschilds' chestnut tree (July 13); smashed into the Trocadéro Hotel (Aug. 8)

1901 *No. 6,* one-person airship

appearance: fat cigar (elongated ellipsoid)

dimensions: 108 feet long

gas capacity: 22,239 cubic feet

motor: 12-horsepower

characteristics: carburetor and oil lubrication system designed to work at any inclination; water-cooled engine; pusher propeller performance: wins Deutsch prize for circling the Eiffel Tower (Oct. 19); extensive ballooning over the Mediterranean; immersed in the Bay of Monaco (Feb. 13, 1902); was knifed in London's Crystal Palace (May 27, 1902); was repaired and shipped to Brooklyn (July 1902) with 115-foot gasbag

1902 *No. 7,* racing airship

appearance: fat cigar

dimensions: 131 feet, 23 feet at greatest diameter

gas capacity: 45,000 cubic feet

motor: 45 horsepower

characteristics: two propellers, fore and aft, driven by a single engine performance: never raced; was knifed at the St. Louis Exposition (June 1904)

1903 *No. 9, Baladeuse,* world's first aerial car
appearance: chubby balloon a third the size of *No. 6*
gas capacity: 7,700 cubic feet
motor: 3 horsepower
performance: 12–15 mph; flew everywhere; transported first child in a powered balloon (June 26); piloted by first woman to fly solo (late June)

1904 *No. 10,* ten-person airship
gas capacity: 80,000 cubic feet
dimensions: 157 feet, 28 feet at greatest diameter
performance: very limited tests; never carried more than one person

1905 *No. 11,* unmanned monoplane glider
performance: barely left water when towed behind speedboat

1905 *No. 12,* helicopter with two propellers
performance: never completed

1905 *No. 13,* airship
characteristics: combined hydrogen gasbag and hot-air generator
performance: never completed

1905 *No. 14,* airship
motor: 14 horsepower
performance: served as an aerial tug to pull heavier-than-air plane (July 19, 1906)

1906 *No. 14-bis, Bird of Prey,* airplane
appearance: canard biplane

dimensions: 40 feet long, 33-foot wingspan
motor: 50 horsepower
performance: flew 37 feet (Sept. 13); 722 feet in 21.2 seconds
   (Nov. 12); first plane flight in Europe; world's first plane
   flight in public

1907 *No. 15*, biplane
   performance: collapsed before leaving ground

1907 *No. 16*, airship/airplane hybrid
   performance: collapsed on the ground

1907 *No. 17*, biplane
   performance: never built

1907 *No. 18*, hydroplane
   performance: never left the water

1907 *No. 19*, prototype of *Demoiselle* sports plane
   appearance: bamboo monoplane
   dimensions: 26 feet long, 18-foot wingspan
   motor: 18 horsepower
   performance: top-heavy because of position of engine above
      pilot's head

1909 *No. 20, Demoiselle,* world's first sports plane
   appearance: a dragonfly with silk-covered wings
   dimensions: similar to *No. 19*
   motor: 18 horsepower
   performance: set 55.8 mph speed record (Sept.); widely copied
      in the United States and Europe

1920s motorized skis to ascend mountains

1920s slingshot-thrown life preserver

1920s treat-dragging device to keep greyhounds moving around
   racetrack

Index